est Coast

Living with the Shore
Series editors, Orrin H. Pilkey and William J. Neal

The Beaches Are Moving: The Drowning of America's Shoreline
New edition
Wallace Kaufman and Orrin H. Pilkey

Living by the Rules of the Sea
David M. Bush, Orrin H. Pilkey, and William J. Neal

Living with the Coast of Alaska
Owen Mason, William J. Neal, and Orrin H. Pilkey

Living with the Alabama-Mississippi Shore
Wayne F. Canis et al.

Living with the California Coast
Gary Griggs and Lauret Savoy et al.

Living with the Chesapeake Bay and Virginia's Ocean Shores
Larry G. Ward et al.

A Moveable Shore: The Fate of the Connecticut Coast
Peter C. Patton and James M. Kent

Living with the East Florida Shore
Orrin H. Pilkey et al.

Living with the West Florida Shore
Larry J. Doyle et al.

Living with the Georgia Shore
Tonya D. Clayton et al.

Living with the Lake Erie Shore
Charles H. Carter et al.

Living with Long Island's South Shore
Larry McCormick et al.

Living with the Louisiana Shore
Joseph T. Kelley et al.

Living with the Coast of Maine
Joseph T. Kelley et al.

Living with the New Jersey Shore
Karl F. Nordstrom et al.

From Currituck to Calabash: Living with North Carolina's Barrier Islands
Second edition
Orrin H. Pilkey et al.

The Pacific Northwest Coast: Living with the Shores of Oregon and Washington
Paul D. Komar

Living with the Puerto Rico Shore
David M. Bush et al.

Living with the Shore of Puget Sound and the Georgia Strait
Thomas A. Terich

Living with the South Carolina Shore
Gered Lennon et al.

Living with the Texas Shore
Robert A. Morton et al.

The Pacific Northwest Coast

Living with the Shores of Oregon and Washington

Paul D. Komar

The Living with the Shore series is funded by
the Federal Emergency Management Agency

Duke University Press Durham and London 1997

Third printing, 2000

Printed in the United States of America
on acid-free paper ∞
Typeset in Minion.
Library of Congress Cataloging-in-
Publication Data appear on the last
printed page of this book.

In memory of William Wick, who as Director
of the Sea Grant College Program at Oregon
State University valued the coast and saw the need
for research that would help in its preservation

Contents

Figures

Preface

When most people think of the Northwest coast, high waves crashing against rocks or surging across wide, sandy beaches come to mind. Much of the attraction of the ocean shores of Oregon and Washington lies in the great variety of the scenery: high, rocky headlands descending precipitously into the sea, sandy beaches fronting sea cliffs, and large expanses of blowing dune sand, all backed by the tree-covered slopes of the Coast Range and the Olympic Mountains. For some, the main attraction of the ocean shores of the Northwest is their seclusion; one can be alone there and in harmony with the forces of nature.

Oregon and Washington are among the fastest growing states in the nation, and their ocean shores are the sites of much of the associated development. More and more sea cliffs and migrating dune fields are occupied by houses, condominiums, motels, and restaurants. Such development imposes a need for permanence on an otherwise dynamic coast. Each winter, storm waves combine with nearshore currents to cut back beaches and cliffs, often resulting in significant damage to coastal properties, both private and public. The response has been to erect shore protection structures, and more and more beaches are backed by massive seawalls and mounds of rock rip-rap. Too often these structures are built with little regard for the dynamic nature of the coast and the processes of ocean waves and currents that continue to reshape the shore.

This book is intended to provide the general public with information about the Northwest coast—its geological setting and the nature of the waves, currents, tides, and other ocean processes that affect it. We will examine the natural responses of beaches and sea cliffs to ocean processes, particularly at spots where these have led to the erosion of coastal properties. Erosion problems are discussed in a general sense and then documented with specific examples: the erosion of Siletz and Nestucca Spits on the Oregon coast during the extreme storms of 1972 and 1978; the unusual processes that existed during the 1982–83 El Niño, which caused erosion along much of the Northwest coast; and the shoreline retreat that has occurred along Cape Shoalwater on the Washington coast, the most massive erosion on any U.S. coast. Over the years I have presented much of this material as luncheon and dinner talks to the general public. The level of the presentation here is much the same—geared toward the general public and illustrated mostly by my own photographs.

It would be easy to become despondent in the face of the rapidly expanding development on the Northwest coast and the proliferation of shore protection structures. Yet, measures that will lead to the preservation of the remaining natural areas are still possible. These will require basic changes in our attitudes toward development and shoreline management policies that incorporate our present understanding of erosion processes; most of all, they will require the support of the public. It is my hope that this book will be of assistance in achieving those goals.

While undertaking research along the Northwest coast during the past 25 years, it has been my pleasure to work with several colleagues at Oregon State University; foremost among them are Professors William G. McDougal, James Good, Robert A. Holman, and Reggie Beach. I wish to take this opportunity to acknowledge the help they have given me over the years. In addition I owe a considerable debt of gratitude to the many students who have worked with me—much of the material in this book comes from their thesis research projects. Much of chapter 5 is based on the Ph.D. dissertation of Dr. Thomas Terich (Terich 1973). A full account of the development of Jump-Off Joe, described in chapter 9, is presented in Sayre and Komar 1988.

The Pacific Northwest Coast

1 A Northwest Coast Perspective

Early explorers of the Northwest coast were impressed by the tremendous variety of its scenery. Today, visitors can still appreciate those qualities. The low, rolling mountains of the Coast Range serve as a backdrop along most of the ocean shores of Washington and Oregon. In the north the tall Olympic Mountains extend right to the shore, as do the Klamath Mountains in southern Oregon and northern California (fig. 1.1). The coast that fronts the mountains is characterized by high cliffs exposed to ocean waves that are slowly cutting away the land (fig. 1.2). The most resistant rocks persist as sea stacks scattered in the surf. Sand and gravel accumulate only in sheltered areas, where they form small pocket beaches on the otherwise rocky landscape.

The longest stretches of beach are found in the lower-lying parts of the coast away from the high mountains of the extreme north and south. The Long Beach Peninsula, which forms a major portion of the Washington coast, extends northward from near the mouth of the Columbia River as a tongue of land that separates the ocean shores from Willapa Bay. It is the largest sand spit in the Northwest. In Oregon, the longest continuous beach extends about 60 miles from Coos Bay northward to Heceta Head near Florence. This beach is backed by the impressive Oregon Dunes, the largest complex of coastal dunes in the United States.

Along the northern half of the Oregon coast there is an interplay between sandy beaches and rocky shores (fig. 1.3). Massive headlands jut out into deep water, their black volcanic rocks resisting the onslaught of even the largest storm waves. Between these headlands are stretches of sandy shore. Portions of these beaches form the ocean shores of sand spits such as Siletz, Netarts, Nehalem, and Bayocean. Landward from most of the spits are estuaries of the rivers that drain the Coast Range. Within each estuary, the fresh water of the river mixes with seawater before passing into the open

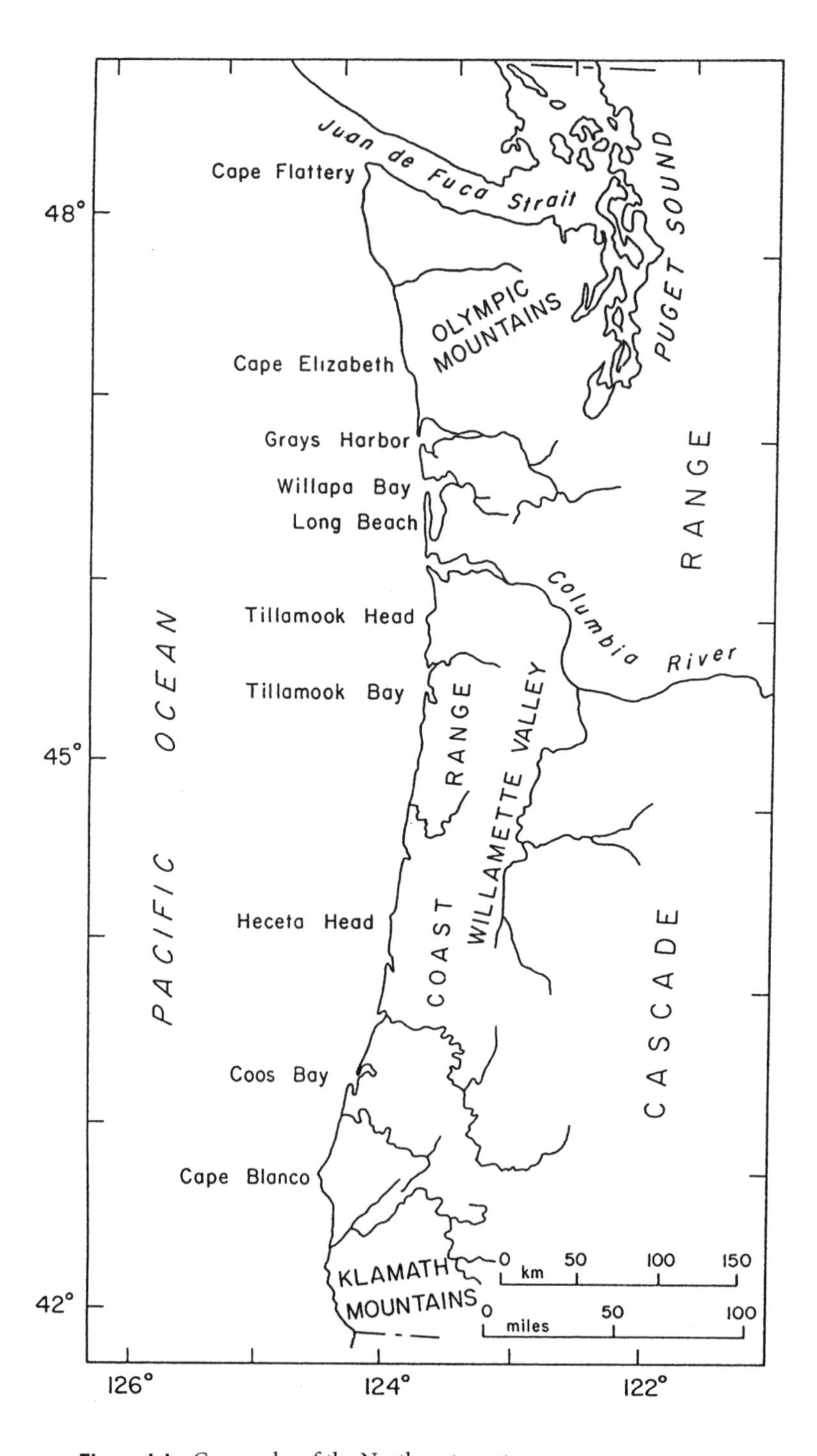

Figure 1.1 Geography of the Northwest coast, the seaward edge of Washington and Oregon.

Figure 1.2a The variety of scenery found along the Northwest coast: *top,* the rocky shores of northern Washington (photo by T. Terich); *bottom,* the flat, sandy beach of the Long Beach Peninsula (photo by T. Terich).

ocean through the narrow inlet at the tip of the spit. Other bodies of water, such as Netarts Bay and the extensive Willapa Bay in Washington, are inland intrusions of the sea that receive little freshwater flow from rivers.

The first Western explorers and settlers were attracted to the Northwest coast by the richness of its natural resources. Earliest to arrive were the traders, who obtained pelts of sea otter and beaver from the Indians. Later came prospectors, who first sought gold in the beach sands and coastal mountains but often were content to settle down and "mine" the fertile farmlands along the river margins. Others turned to fishing, supporting themselves by harvesting the abundant Dungeness crabs, salmon, and other fish in the coastal waters. Also important to the early economy of the

Figure 1.2b *Top,* Netarts Sand Spit in northern Oregon; *bottom,* Heceta Head on the mid-Oregon coast.

coast—and today's economy, too—were the vast tracts of cedar and Sitka spruce. Today, however, the most important "commodity" for the Northwest coast economy is the vacation visitor. Vacationers arrive in thousands during the summer months, but it is still possible to leave Highway 101 and find a secluded beach or a quiet forest trail.

There is cause for concern that the very qualities that have made the Northwest coast so attractive to people are being destroyed by human activities. Like most coastal areas, the Northwest is experiencing the pressures of development. Homes and condominiums are being constructed immediately behind the beaches, within dunes and atop cliffs overlooking the ocean. Everyone wants a view of the waves, the passing whales, and the evening sunset, as well as easy access to a beach. These desires are not always compatible with nature, however, and more and more homes are threatened by and sometimes lost to beach erosion and cliff landsliding.

Problems can usually be avoided if people will only recognize that the coastal zone is fundamentally different from inland areas and act accord-

Figure 1.2c *Top,* the massive dunes of the central Oregon coast; *bottom,* the rocky coast typical of southern Oregon.

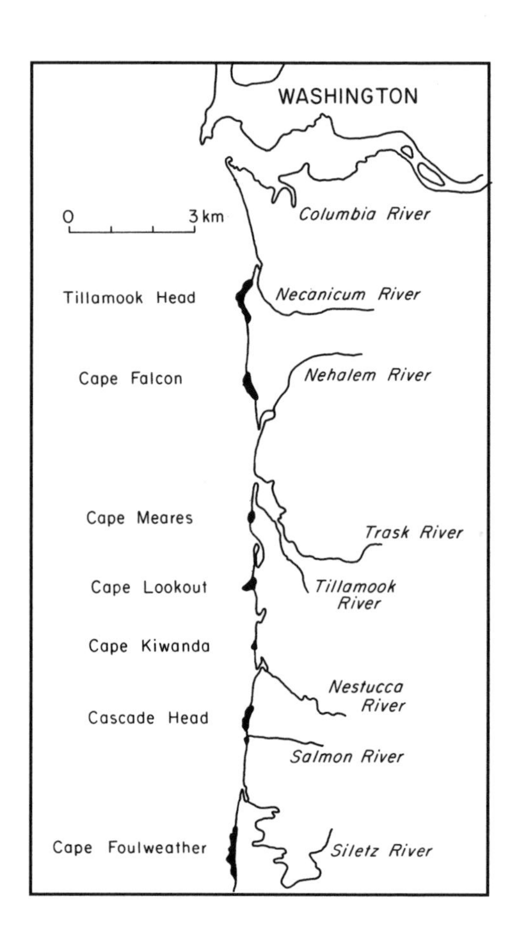

Figure 1.3 The northern half of the Oregon coast, with large, rocky headlands separated by stretches of sandy beach.

ingly. This requires some knowledge of ocean waves and currents and how they shape beaches and attack coastal properties. It also requires an understanding that some parts of the shore are unstable, subject to erosion and sudden landslides. Familiarity with the processes and types of problems experienced in the past can be helpful in selecting a safe homesite. It can also enhance your enjoyment of the coast and lead to an appreciation of the qualities of the Northwest coast that can be and must be preserved.

2 Geological Evolution of the Northwest Coast

The Northwest coast has been shaped by the interplay between geologic, climatic, and oceanographic processes. Tectonic activity deep within the earth accounts for the wide variety of rocks found along the shores of Oregon and Washington—mudstones and sandstones that originated in the sea as well as hard, black rocks that formed in volcanoes. Their differing degrees of resistance to erosion are responsible for the coast's morphology. Sand and gravel eroded from the rocks have been shaped by waves into beaches and blown inland to form large tracts of dunes. Also important to the development of the present-day coast was the rise and fall of the sea level during the ice ages. Those sea level cycles, superimposed on a progressive long-term rise of the land resulting from continued tectonic activity, account for many of the landscape details visible today on the Northwest coast.

Plate Tectonics and Continental Growth

The geological setting of the Northwest has had much to do with the development of the physical features of Oregon and Washington. Indeed, the very existence of the land mass forming those states can be credited to the extraordinary geological processes of plate tectonics.

Beginning in the 1960s, scientists began to develop a fuller appreciation of the impermanence and mobility of the earth's surface. The science of plate tectonics arose primarily from discoveries within the ocean basins. Oceanographers found large mountains in the depths that form essentially continuous ridges and rises around the earth—a gigantic, globe-encircling system that is some 60,000 miles long. Sampling dredges bring up fresh volcanic rocks from the crests of these ridges, and there is virtually no sediment cover, indicating that the ridge system is very young in geological terms. The ridge crest is cut along its length by a steep-walled, symmetrical

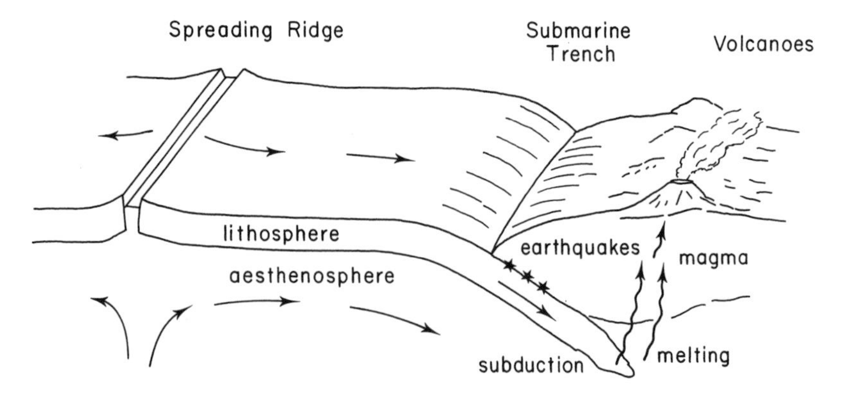

Figure 2.1 The formation of ocean crust at a spreading ridge and its subduction where the oceanic plate collides with a continental mass. Stars denote earthquakes formed by the plates scraping together. The melting of the subducted plate gives rise to volcanoes.

canyon thousands of feet deep. Abnormally large amounts of heat flow from the earth's interior in this area, which is a zone of frequent earthquakes.

These features are all associated with the formation of new ocean crust at the ridges. The earthquakes occur when the crust fractures as it is pulled apart, and the heat is associated with molten lava rising from deep within the earth to form the new crust.

Such evidence from the ocean basins led to the concept of seafloor spreading, the notion that new ocean crust is formed at the submarine ridges by fracturing along the crest, with molten rock rising to fill in the gap. Once the new crust solidifies, it splits roughly in half and then spreads away from the ridge, the two halves moving down opposite flanks into deeper water. Figure 2.1 illustrates the process. The conditions depicted on the figure are those of the seafloor west of South America, but they are similar to those found off the Northwest coast. The new crust formed at the spreading ridge moves slowly away from the ridge (shown by the arrows in the figure) as it is displaced by still newer crust. The formation of new crust at ocean ridges might seem to imply that the earth is expanding and slowly increasing its surface area. This would be the case if it were not for the fact that older crust is simultaneously destroyed in a process called subduction (fig. 2.1). The downbuckling of the ocean crust forms a submarine trench at the continental margin, and the site where this slab of descending ocean crust scrapes against the overlying continental mass is another earthquake zone. The subducting crust eventually gets deep enough within the earth to be remelted, and this molten rock supplies the lava that emerges in the numerous volcanoes of the Andes Mountains of South America and the Cascades of the Northwest.

This basic pattern occurs around the entire margin of the Pacific Ocean, which has come to be known as the "Ring of Fire" as a result. In some areas the spreading ocean crust is subducted beneath another segment of ocean crust. In that case, a trench forms and the volcanoes develop into an island chain with an arcuate shape (arcuate because it lies on a spherical earth); examples of such island arcs are the Aleutian Islands of Alaska and the Mariana Islands of the western Pacific.

The concept of seafloor spreading, forming and then destroying ocean crust, has been firmly established by many lines of evidence. Its ultimate cause is believed to be the slow movement of rock within the interior of the earth, rock that is sufficiently hot to make it soft and plastic but not fully molten. The ocean crust and the continents are rafted along by the flow, which apparently is part of a system of convection currents heated at still greater depths within the earth. Being less dense but thicker than the ocean crust, the continents float higher and extend down deeper, much as an iceberg floats on water. The picture of the earth that emerges is of a mobile surface with impermanent continents and oceans. Nothing seems to be fixed. The crustal plates move about, colliding to form mountains or being subducted into the earth's interior.

These processes of global tectonics account for the principal features of the Northwest (Drake 1982; Orr et al. 1992). Mount Rainier, Mount Saint Helens, Mount Hood, and the other large peaks of the Cascades are volcanoes formed from oceanic rocks that melted during subduction and descended into the interior of the earth. The submarine trench off the Northwest coast is not nearly as well developed as the one off the coast of South America shown in figure 2.1, and it is so filled with ocean sediments that it is not apparent in the bathymetry of the seabed.

The spreading ridges off the Northwest coast—the Juan de Fuca and Gorda Ridges (fig. 2.2)—are offset by the Blanco Fracture Zone, a giant fault in the seabed that has been the source of a number of earthquakes felt along our coast. The two segments of the ridge formed by the fracture zone often act as crustal plates—respectively, the Juan de Fuca plate and the Gorda plate. New ocean crust is being formed at the Juan de Fuca and Gorda Ridges, and its general movement is eastward toward the continent, which sits atop the North American plate. The spread of the Juan de Fuca Ridge and the consequent movement of ocean crust toward the continent is estimated to occur at a rate of 2.3 inches per year (T. Atwater 1970). However, that is not the rate at which this ocean crust collides with North America, because the latter mass is also moving. The actual convergence between the ocean crust and the land mass is only about 1 inch per year and is at an oblique angle rather than being head-on.

Subduction of the ocean plate beneath South America has produced catastrophic earthquakes in Peru and Chile (fig. 2.1). Until recently, scien-

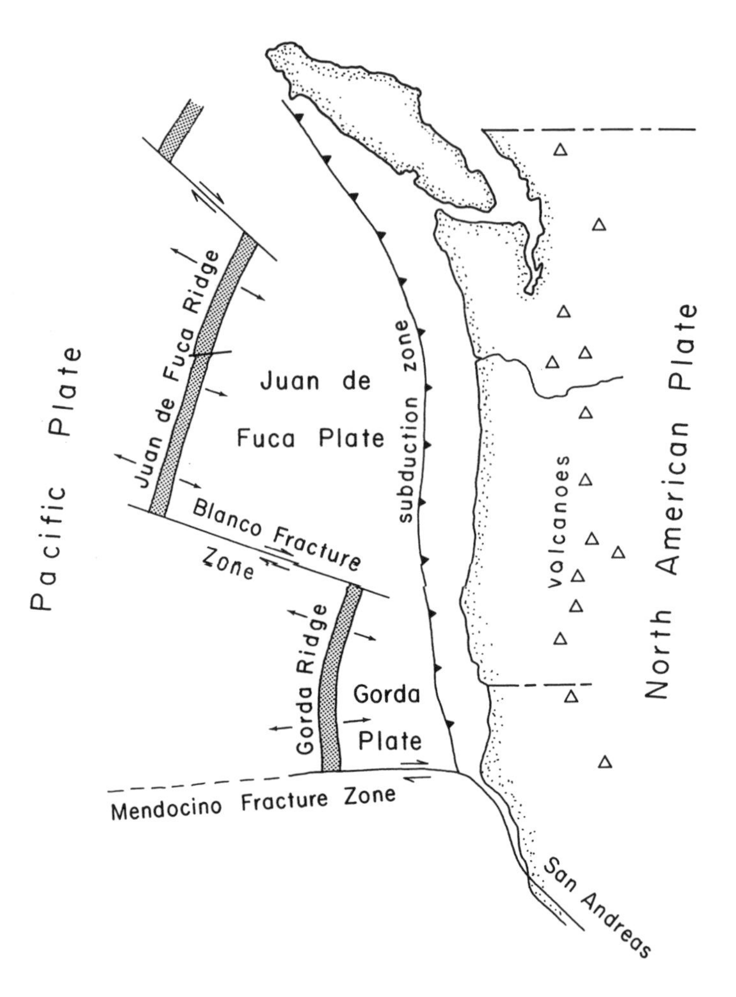

Figure 2.2 The Juan de Fuca and Gorda plates beneath the sea west of Oregon and Washington move toward the coast as new crust is formed at the ocean ridges. The subduction zone is offshore, seaward of the continental shelf.

tists were puzzled by the absence of such subduction earthquakes in the Northwest. Now, however, there is strong evidence that such earthquakes have occurred, but at intervals of hundreds to thousands of years. None has taken place within the last two or three centuries, since Europeans came to the Northwest, although myths and archaeological evidence indicate that the native inhabitants witnessed past catastrophic events (Hall and Radosevich 1995).

The principal evidence for major prehistoric earthquakes associated with subduction comes from investigations of estuarine marsh sediments buried by sand layers. These deposits suggest that portions of the coast subsided abruptly and were then overwashed by extreme tsunamis, or "tidal waves," that swept over the area and deposited the sand (B. F. Atwater 1987; Darienzo and Peterson 1990; B. F. Atwater and Yamaguchi 1991; Darienzo et al. 1994; Peterson and Priest 1995). The number of these layers found in Willapa Bay and Netarts Bay and carbon-14 dating of the sediments indicate that catastrophic earthquakes have occurred along the Northwest coast at least six times in the past 7,000 years, at intervals ranging from 300 to 600 years, with the last having occurred about 300 years ago. Given this estimated frequency of occurrence, there is a strong possibility that another major subduction earthquake and tsunami will occur during the next 100 years.

The lack of subduction earthquakes within historic times suggests that the ocean plates and the North American plate are temporarily locked together. If so, they must be accumulating energy, much as a compressed spring does. This is an ominous situation. Ideally, small earthquakes periodically release the tension generated when two plates come together, allowing the plates to slide by one another. The longer the plates remain locked together, the greater the amount of energy stored, and the more catastrophic the resulting earthquake when the plates finally do break apart along the subduction zone.

Scientists are still debating the magnitudes of past subduction earthquakes in the Northwest, and hence the magnitude of an earthquake that could be expected in the future. The subduction process has caused considerable deformation and fracturing of the upper continental plate, and some scientists have suggested that this deformation releases part of the accumulating strain, thus reducing the magnitude of subduction earthquakes (Goldfinger et al. 1992; McCaffrey and Goldfinger 1995). Even so, they estimate that a future earthquake could be on the order of magnitude 8, comparable to the earthquake that destroyed San Francisco in 1906. Estimates of the magnitudes of past earthquakes, based on the along-coast lengths of ruptures in the subduction zone inferred from the simultaneous subsidence of marshes and accumulations of tsunami sands within several estuaries (Darienzo and Peterson 1995), indicate that at least some had magnitudes greater than 8, possibly greater than 9.

Perhaps the most definitive evidence for subduction earthquakes and tsunamis on the coast of Oregon and Washington has come, surprisingly, from Japan (Satake et al. 1996). In addition to devastating our coast, large tsunami waves generated off the Northwest coast would also travel across the Pacific Ocean and eventually reach the shores of Japan, though they would have weakened by then. The Japanese have kept detailed records on damage-causing tsunamis for nearly 400 years. These records reveal that damage occurred at a number of sites along the coast of Japan in January 1700, resulting from a tsunami that could only have come from the northwestern United States. Considering that it would have taken about 10 hours for the tsunami waves to travel across the Pacific, the Japanese estimated that the subduction earthquake must have occurred at about 9:00 P.M. on January 26, 1700, a time that is consistent with Native American legends, which say the quake occurred on a winter night.

The evidence is conclusive: major subduction earthquakes have occurred in the Northwest, causing large portions of the coast to drop immediately by several feet and generating huge tsunamis that washed over our coast and traveled in the other direction as far as Japan.

Sediments begin to accumulate on ocean plates as soon as the plates are formed, the thickness increasing with time. The Juan de Fuca and Gorda plates have unusually thick accumulations of sediments due to their close proximity to the continent. (Rivers and coastal erosion deliver large amounts of mud and sand to the ocean.) Much of the accumulated sediment is scraped off during subduction and is added to the continental mass, leading to long-term growth. Subduction also causes the progressive uplift of the land mass and carries deepwater sediments and the fossils they contain up into relatively shallow water. Remains of organisms that once lived at depths of 2,000–3,000 feet have been found in sediments on the Oregon continental shelf and slope at depths less than 600 feet (Byrne et al. 1966; Kulm and Fowler 1974), clear evidence that significant uplift of the Northwest margin has taken place.

Nearly all of Oregon and Washington was created by the accretion to the continent of ocean sediments and a series of volcanic island arcs similar to the present-day Aleutian Islands. The Blue Mountains of easternmost Oregon are thought to be portions of ancient ocean crust and island arcs that were added to the continent during the late Triassic and Cretaceous geologic periods (Brooks 1979)—the age of dinosaurs—150–250 million years ago. Before that, what we now think of as the Northwest was part of a deep ocean basin.

The oldest rocks found in the Coast Range of western Washington and Oregon date back to the Paleocene and Eocene epochs, about 40–60 million years ago. These old rocks are ocean basalts much like those being formed today by volcanic activity at spreading ridges. Apparently, a ridge similar to the present-day Juan de Fuca Ridge was situated to the west. The Klamath Mountains were present to the south, but otherwise there was an embayment in the coast with deep ocean water over the areas that are now the Coast Range, the Willamette Valley of Oregon, and the Puget Lowland of Washington. A center of unusually strong volcanic activity on the ridge crest formed a chain of islands that accreted to the continent and are now present as a north-south series of peaks in the Coast Range.

In the Oligocene epoch, some 30 million years ago, thick layers of melted rock intruded into the older ocean crust rocks that now form the base of the Coast Range. These intruded volcanic rocks were extremely resistant to erosion and today form the higher peaks of the Coast Range; Marys Peak west of Corvallis, the highest of the Coast Range mountains (elevation 4,097 feet), is an example.

During the Miocene, 30–35 million years ago, volcanic activity generated the immense flows of the Columbia River basalts (Orr et al. 1992). Volcanic activity recurred to the west at the same time. This activity was somehow connected with the generation of the Columbia River basalts because the rocks are almost exactly the same. These Miocene volcanic rocks are particularly important to the morphology of the modern coast because they are resistant to wave attack and form many of the major headlands along the Oregon coast; Yaquina Head, Cape Foulweather, and Cape Lookout are examples (fig. 2.3).

Figure 2.3 Cape Lookout on the north Oregon coast is composed of ancient volcanic rocks highly resistant to wave attack. From the Oregon Highway Department.

The ocean embayment into western Oregon and Washington persisted for many millions of years, accumulating seafloor sediments all the while. A major river entered the embayment from the southeast, carrying large loads of sediment that built out a delta (Chan and Dott 1983). Masses of this sediment periodically slumped off the delta and flowed into the deep water of the embayment, settling out of suspension to form sand layers that are graded with the coarsest sand particles at the base and finest at the top (Heller and Dickinson 1985). These layers of deep-sea sands were eventually uplifted onto the land and are now visible in road cuts of highways crossing the Coast Range. Swamps developed along the margins of the ocean embayment, and the slow accumulation of dead marsh plants produced the coal beds found in the Coos Bay area.

The embayment persisted until the Pliocene epoch, about 5 million years ago, at which time the Coast Range began to emerge from the sea and western Oregon and Washington came into being. In terms of geologic time, the Northwest is thus very young. With the land's emergence from the sea, rain, rivers, and ocean waves went to work to erode what formerly had been deep-sea crustal rocks and volcanoes, and ocean sediments hardened into rock. These processes have etched out the land, cutting away the weaker rocks and leaving behind the more resistant ones to form the peaks of the Coast Range and the headlands along the coast. The modern morphology of the coast is the product of this erosion over the last 5 million years.

Sea Level and Its Imprint on the Northwest Coast

The most recent epoch of the earth's history, roughly the last two million years, has seen the repeated advance and retreat of glaciers over the continents. At their maxima, the glaciers moved down across Canada and into the northern portions of the United States. The water that formed the ice came from the oceans, and each build-up and advance of the glaciers lowered the sea level, causing the shoreline to migrate out to what is now relatively deep water. Because the water was alternately locked up in glaciers and released, the level of the sea went through cycles of highs and lows, and this, too, has had a major impact on the morphology of the Northwest coast.

The continental glaciers moving down from the north only just reached the edge of the Northwest coast. The ice advanced as a tongue into the Puget Lowland, reaching south of the present-day site of Olympia, Washington (fig. 2.4). Glaciers covered all the hills of the Puget Lowland and the San Juan Islands and lay high against the flanks of the Olympic and Cascade Mountains. Another tongue flowed seaward and carved out the trough that is now the Juan de Fuca Strait (fig. 2.4). Isolated glaciers formed in the

Olympic and Cascade Mountains, and small remnants still remain there as evidence of the ice ages.

Glaciers advanced into the Puget Lowland at least four times at intervals of hundreds of thousands of years. During the last advance the glacier acted as a dam at the north end of the Puget Lowland, trapping water to form a large lake. This lake filled the entire lowland, bounded by the ice wall on the north and by mountains on the other three sides. The water level rose until it spilled over the lowest pass of the Coast Range and then flowed down what is today the Chehalis River, which drains into Grays Harbor. This explains why the modern Chehalis River, which is relatively small, occupies such a large valley. When the Chehalis drained the glacial lake of the Puget Lowland, its discharge of water was probably several times greater than that of the modern Columbia River (McKee 1972).

The Columbia River also saw floods of immense proportions during the ice ages, and its discharge was also considerably greater than at present (Bretz 1969; Baker 1973; Allen et al. 1986). The Columbia floods also originated in an ice-dammed lake, this one located in the area of Missoula, Mon-

Figure 2.4 The maximum advance of glaciers into Washington formed a lobe in the Puget Lowland and another that flowed seaward and cut the Juan de Fuca Strait. After McKee 1972.

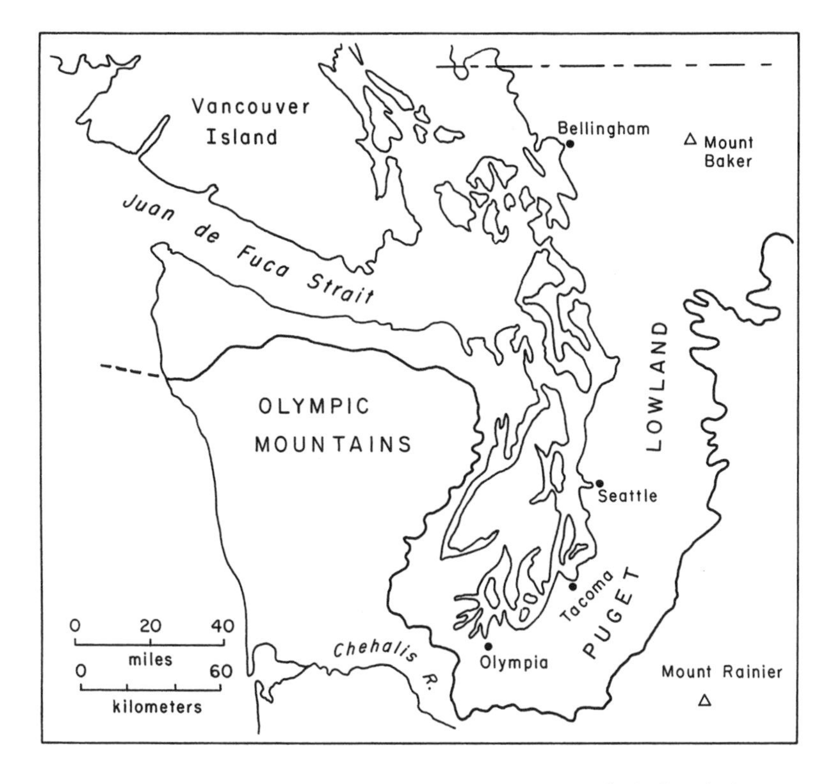

tana. But this lake's water was released suddenly when the ice dam failed, producing catastrophic floods. The released water flowed out across eastern Washington, carving large channels into even the most resistant rocks. The floodwaters funneled into the Columbia River, but its channel could not hold such large volumes of water, and part of it spilled up into the Willamette Valley, forming a temporary lake.

Since the glaciers did not actually reach most of the Northwest coast, their main impact on the coastal morphology has been indirect and is mostly the result of the changes in the level of the sea produced by the alternating growth and melting of the ice sheets. During high stands of the sea hundreds of thousands of years ago, the waves eroded the rocks and beveled them flat to form nearly level terraces. Those ancient terraces were carried upward by tectonic activity and can now be found miles inland and at high elevations in the Coast Range.

Most apparent of the uplifted marine terraces is the youngest that backs the present-day shoreline along many miles of the coast (fig. 2.5). This terrace is young enough to give the appearance of having recently emerged from the sea. Cliffs cut into the terrace by waves reveal that its lower portions consist of ancient mudstones eroded flat by waves thousands of years ago. A uniform layer of light-colored sand occurs in the sea cliff above the gray mudstone. Close study of the sand reveals that much of it was deposited on ancient beaches—its layering and other features are identical with those found in modern beaches. Dune sands and old soil horizons are also evident in the terrace deposits, further illustrating that the environment of that ancient coast was substantially the same as today's.

The lowermost uplifted terrace is the youngest of a series of marine terraces that in places form stairways up the flanks of the Coast Range. The highest and oldest terrace in the series reaches elevations up to 1,600 feet in southern Oregon (Baldwin 1945). The older the terrace, the more degraded it has become through erosion, and also the more warped by tectonic activity, so that old terraces are no longer level platforms (Kelsey 1990). The beach and dune sands of these old terraces are weathered and cemented by rusty iron oxide. However, the sand in even the oldest terraces is still identifiable as having originated in beaches and dunes. Some terrace sands have accumulations of valuable black sands like those found on the modern beach. These ancient black sands have been mined for gold, and excavations within the deposits have exposed drift logs that were deposited on beaches more than 100,000 years ago.

The presence of marine terraces high up on the western slopes of the Coast Range does not mean that sea levels were formerly that much higher than at present. They are there because the coastal margin of the Northwest has been tectonically rising for hundreds of thousands of years. This progressive rise of the land, together with the global oscillations in the absolute

Figure 2.5 *Top,* the marine terrace at Otter Rock, Oregon, showing the gray Tertiary mudstones dipping seaward beneath the lighter-colored Pleistocene sand; *bottom,* close-up of the Pleistocene sand showing layers typical of beach deposits and an 80,000-year-old drift log.

level of the sea due to glacial advances and retreats, created the stairways of terraces. Each terrace does record a high stand of sea level, but not necessarily a level higher than at present. All the high stands are probably not recorded because some terraces may have been eroded away, just as the sea is presently eroding the youngest uplifted terrace.

Our knowledge of the ages of the various marine terraces is limited by the scarcity of fossils that can be used to date them. Fossil shells and corals found in the lowermost terrace in southern Oregon lived about 80,000 years ago (Kelsey 1990; Muhs et al. 1990), which is fairly conclusive evidence that the lowest terrace correlates with the high stand of sea level that reached a maximum 80,000 years ago. Based on that date plus the known ages of still earlier high stands of sea levels and an estimated rate of uplift for the Northwest coast, the ages of the next two higher terraces have been placed at about 105,000 and 125,000 years before the present (Adams 1984; Kelsey 1990; Muhs et al. 1990). This points to the considerable antiquity of the oldest terraces that have been lifted high up onto the slopes of the Coast Range.

The tectonic rise of the marine terraces, and of the Northwest coast in general, has not been solely vertical. The uplift has also been part of a rotation about an inland pivot line located somewhere within the central lowlands of Puget Sound and the Willamette Valley (fig. 2.6; Reilinger and Adams 1982; Adams 1984; Kelsey 1990). The farther west from that pivot line (i.e., the closer to the coast), the more rapid the local vertical uplift. The rotational uplift of the coast has altered the slopes of the marine terraces, and the older the terrace, the more it has been affected. Some terraces now actually slope landward rather than seaward (Adams 1984). The uplift rotation is not uniform north and south along the coast because some areas are rotating upward faster than others. Surveys along a north-south line extending the full length of the Oregon coast (fig. 2.7) demonstrate that the smallest rates of uplift are occurring on the central coast between Newport and Tillamook, with progressively higher rates to the south toward the California border and to the north toward Astoria (Vincent 1989; Komar and Shih 1993; Mitchell et al. 1994).

According to the survey results shown in figure 2.7, the southern half of the Oregon coast is rising faster than the global rate of sea level rise, and most of the northern half is being submerged by the rising sea. The submergence rates of the north Oregon coast are on the order of 1–2 millimeters per year (4–8 inches per century), which is much smaller than the 4–6 millimeters per year (16–24 inches per century) common along the east and Gulf

Figure 2.6 The uplift of the coast due to the rotation of the land about a pivot point or line located in the central valley of the Puget Lowland and Willamette Valley.

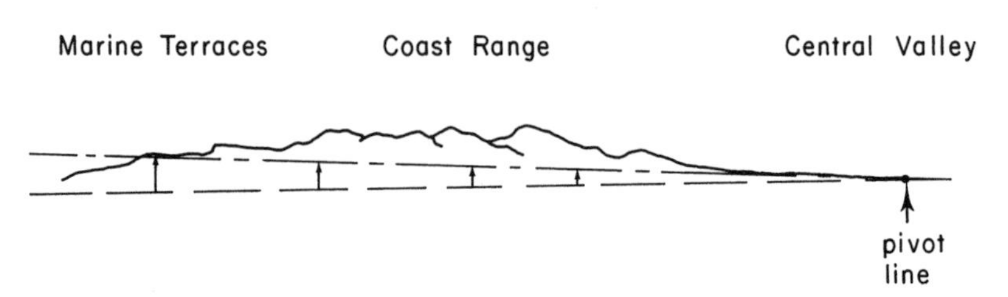

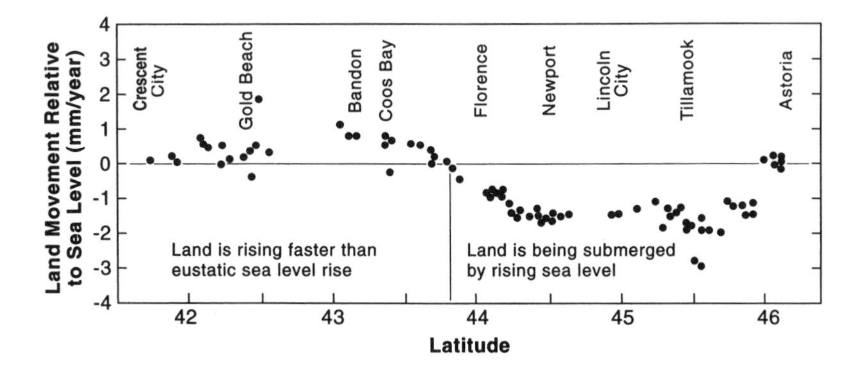

Figure 2.7 Elevation changes along the length of the Oregon coast from Crescent City in northern California north to Astoria on the Columbia River, derived from P. Vincent's analyses of geodetic surveys (Vincent 1989). Elevation changes relative to the varying sea level, with positive values representing a rise in the land relative to the global rise in sea level, and negative values the progressive submergence of the land by the sea (1 inch ≈ 25 millimeters).

coasts of the United States. The uplift of the Northwest coast measured during historic times (the last 100 years) is interpreted as being the result of accumulated strain from subduction of the oceanic Juan de Fuca plate beneath the continent and has probably been under way since the last subduction earthquake occurred in 1700.

The heights of the marine terraces along the coast must reflect the long-term difference between the uplift that occurs during periods between earthquakes and the abrupt lowering that occurs during the earthquakes. For example, the marine terrace at Bandon on the southern Oregon coast formed 80,000 years ago. Its uplift rate during historic times has been about 2 millimeters per year (8 inches per century), so one might expect the terrace to have risen more than 500 feet since its initial formation. In fact, its elevation is only about 100 feet. The discrepancy is likely accounted for by subsidence accompanying seismic events. If the subsidence averages 20–40 inches during each event, then the recurrence intervals would have to be on the order of 300–600 years (Komar et al. 1991), a figure in agreement with the time intervals determined by dating marsh deposits.

Coastal subsidence during an earthquake is probably very uneven; there is considerable local folding and faulting, and this affects the elevations of the terraces. Figure 2.8 shows elevations of the youngest terrace in southern Oregon. The terrace is highest in the Cape Blanco area, which therefore must have the greatest rate of net uplift over the long term (Baldwin 1945; Kelsey 1990; Goldfinger et al. 1992). The height of the terrace decreases markedly toward the north, reaching a low point north of Coos Bay. This points to the significance of the uplift rate in controlling the present-day coastal morphology; the high, wave-cut cliffs of Cape Blanco are in marked contrast to the low area between Coos Bay and Heceta Head, where the sands of the Oregon Dunes have accumulated.

Any records that remain of low stands in sea levels are beneath the sea on the continental shelf. This makes it difficult to learn much about what the Northwest coast must have been like at that time. Drowned beaches and wave-eroded terraces have been identified on many continental shelves, some of them at water depths greater than 300 feet. Dredges have brought up the teeth of mastodons from these depths, evidence of the life that must have flourished long ago when sea levels were lower. However, more important than mastodon teeth has been the dredging up of intertidal organisms and marsh peat, material that when living had a close relationship with the level of the sea. By dating that material we can reconstruct former water levels and determine how much and how rapidly the sea has risen since its most recent low 20,000 years ago (fig. 2.9). At the time of the last glacial

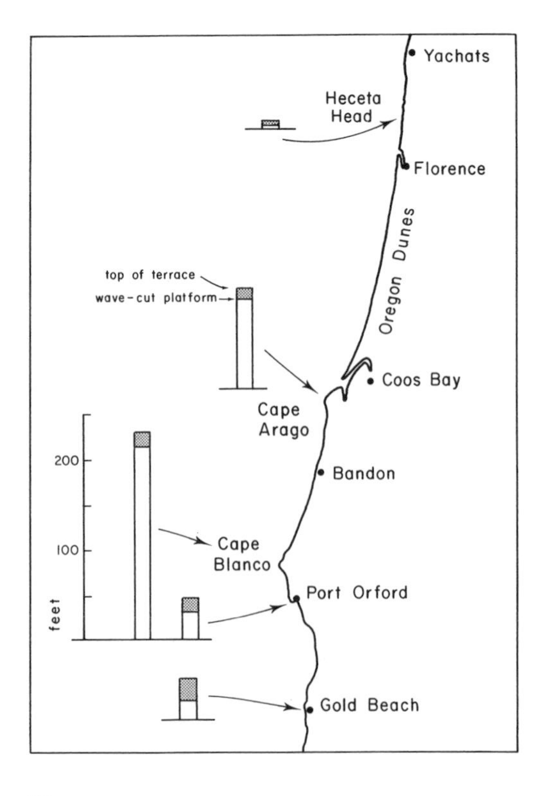

Figure 2.8 Elevations of the youngest marine terrace near Cape Blanco on the Oregon coast. After Baldwin 1945.

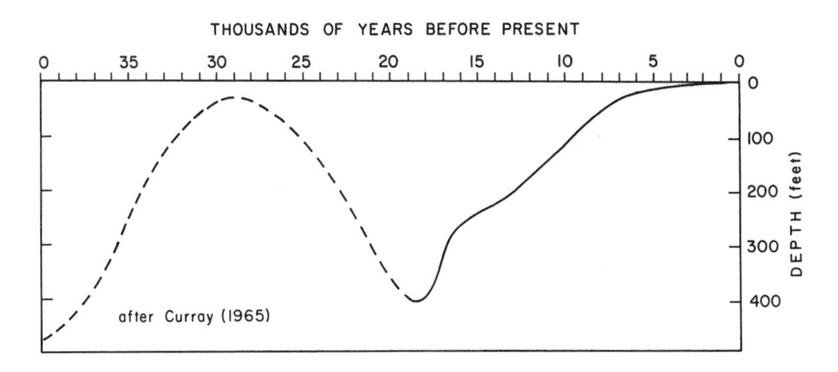

Figure 2.9 Changing sea levels over the past 40,000 years due to the growth and melting of continental glaciers. After Curray 1965.

maximum, the sea was nearly 400 feet lower than it is now. As the glaciers melted, the sea rose rapidly until about 6,000 years ago, when the rate slowed. The dashed line in figure 2.9 is an estimation of the sea level prior to 20,000 years ago, during the cycle associated with the previous glacial advance and retreat. Glacial fluctuations and sea level cycles have been occurring for more than a million years, with numerous high stands and low stands. The sea has repeatedly passed across the continental shelf, and the shelf itself is largely the product of the resulting erosion. Unfortunately, the shelf sediments record only the most recent cycle, that cycle having disrupted or covered most of the deposits laid down by previous cycles of water level change.

The sea continued to rise until about 2,000 years ago (fig. 2.10), when it reached approximately its present level. However, analyses of tide gauge records indicate that the sea is still rising. Tide gauges record the hourly changes in the level of the sea during regular tidal cycles. If we take the measurements recorded on a tide gauge during one day and average them, we obtain the mean water level for that day. If those daily averages are in turn averaged over an entire year, we have a measure of the mean sea level at that tide gauge. When such analyses are repeated year after year, we can monitor the progressive change in sea level that might be caused by the continued melting of glaciers. Data from nearly all the tide gauges that have been in operation sufficiently long to reveal net changes in sea levels have been similarly analyzed (Gornitz et al. 1982; Hicks et al. 1983; Barnett 1984). Figure 2.11 shows 80-year records for four locations. Each location experienced considerable fluctuations in the level of the sea from year to year, with many small ups and downs.

The sea level in any given year is affected by many oceanic and atmospheric processes (see chapter 3), which produce the small fluctuations. In spite of such irregularities, most tide gauge records reveal a long-term rise in the sea that can in part be attributed to glacial melting. The record from New York City (fig. 2.11), which shows a long-term average rise of 3.0 millimeters per year (12 inches per century), is typical. The record from Galveston, Texas, also shows a rise in sea level, but the average rate is much higher, 6.0 millimeters per year (24 inches per century). Of course the actual level of the sea is not rising faster at Galveston than at New York City. The difference is the result of changes occurring on land. The Galveston area is subsiding as a result of the pumping of oil and groundwater, and the higher rate in water level rise registered on that tide gauge represents the combined effects of the local land subsidence and the actual rise in sea level.

In contrast, Juneau, Alaska, is rising faster than the global sea level, and its tide gauge record therefore indicates a net fall in the water level relative to the land (fig. 2.11). The record from the tide gauge at Astoria, Oregon, indicates that the level of the sea there has remained relatively constant with respect to the land. During at least the last half century, Astoria has been rising at just about the same rate as the sea. In fact, a careful analysis of the data from the Astoria gauge shows that the land is actually rising slightly faster than the water, the net rate being relatively small; it would amount to only 10–20 millimeters (less than an inch) if it continued for 100 years. The rate derived from the Astoria tide gauge is in agreement with the geodetic survey data analyzed in figure 2.7, which provides a more complete picture of the relative rise of the sea level along the Northwest coast.

Obviously, with different areas of the land moving up and down at different rates, it is difficult to determine how much of the change is actually due to the long-term global rise in sea level associated with glacial melting.

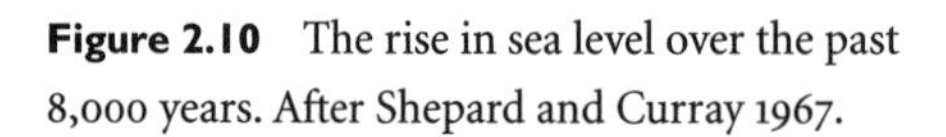

Figure 2.10 The rise in sea level over the past 8,000 years. After Shepard and Curray 1967.

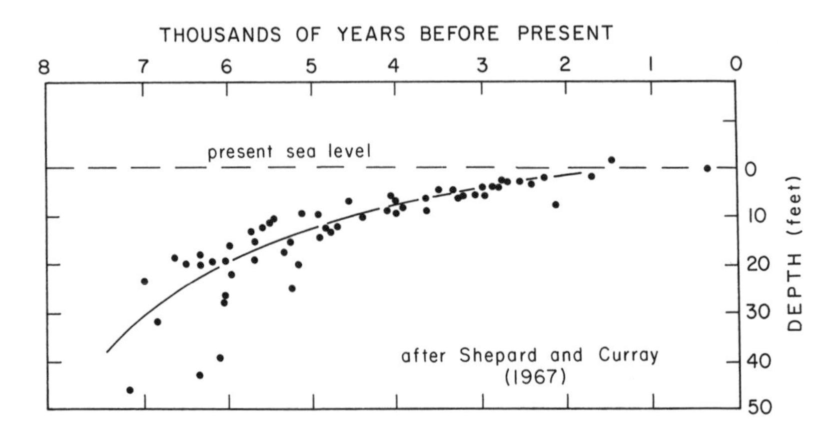

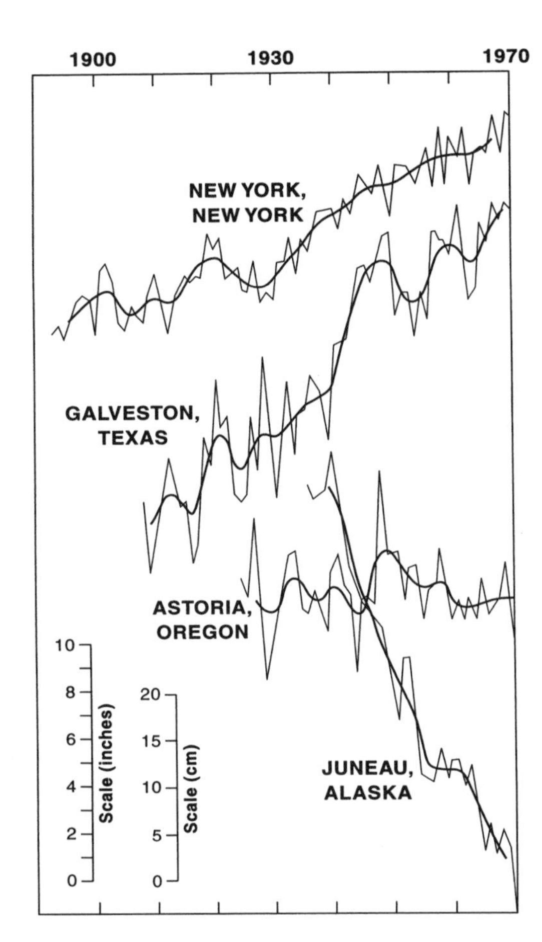

Figure 2.11 The yearly changes in sea level determined from coastal tide gauges. The change at a specific station is the sum of the variations in the actual worldwide sea level and the changing level of the land at that site, and therefore differs from one station to the next (1 inch = 2.54 centimeters). After Hicks 1972.

The best that can be done is to eliminate the records from areas that are undergoing obvious local subsidence or emergence and then combine the remaining records to obtain a worldwide average. The results indicate that sea level is rising between 1 and 2 millimeters per year (Hicks 1978; Barnett 1984). This may not seem like much—it amounts to only about 4–8 inches in a century—but even that small rise can have a significant impact on low-lying coasts.

When areas are also subsiding, it is the net local change that is important to beach erosion and the movement of the sea over the land. A rise of 4 millimeters per year is common on the east coast of the United States, and this rate represents a rise in water level relative to land of 16 inches in a century.

We are fortunate in the Northwest because the level of the sea is presently dropping with respect to the land or is only slightly higher than the tectonic rise (fig. 2.7). The erosion along our coast is much less than it would be if the sea were moving up over the land at the rates found on the

East and Gulf Coasts. However, this advantageous situation may change in the future. A global warming trend has been under way since the 1960s, perhaps a "greenhouse effect" caused by the carbon dioxide and other gases human activities have pumped into the atmosphere. If this warming continues, sea levels are expected to rise faster than at present and could reach levels 2–10 feet higher by the year 2100 (Hoffman et al. 1983; National Research Council 1985). However, the significance of greenhouse warming is still being debated by scientists, and predictions of an associated major sea level rise are uncertain. If there is an accelerated rise in sea level, the greatest impacts would be in low-lying coastal areas of the eastern United States. The effect on the Northwest coast would be smaller and would come later, but we cannot ignore this potential hazard.

The Evolution of the Northwest Coast since the Ice Ages

Twenty thousand years ago, during the last advance of the glaciers, the shoreline of the Northwest was some 20–30 miles west of its present position (fig. 2.12). The beaches were close to the edge of the continent, and a wide, nearly level plain separated the mountains from the ocean. Rivers flowing out of the Coast Range crossed this flat coastal plain at a slow, meandering rate. Today that plain is under water and forms the continental shelf. Portions of the old channel of the Columbia River found on the seafloor indicate that its sediments were delivered to Astoria Canyon, which notches the edge of the continental shelf and slope (fig. 2.12). The river sediments dumped into Astoria Canyon then slumped into the deep ocean basin.

During the lowest stand of sea level, the coastline was probably irregular (fig. 2.12) because the shoreline was close to the shelf edge, which is also irregular. There may have been only pocket beaches, and these probably were of limited extent because of the tendency for the sand to move offshore over the steep bottom slopes. As the glaciers began to melt and the sea level began to rise, the shoreline migrated back across the coastal plain (the present continental shelf). The beaches were relatively continuous, with few if any stretches of rocky shore. No headlands were present on the shores of Oregon and Washington then, and waves transported sand north and south along the ancient beaches, a transport that is not possible today because of the numerous headlands.

Analyses of the minerals in sands from the former coastal plain beaches, which now lie underwater on the continental shelf, indicate that the net movement of sand along those ancient shorelines was to the north (Scheidegger et al. 1971). Most of the beach sand on the Northwest coast consists of grains of quartz and feldspar. Particles of these minerals are transparent or light tan, and this is the color of most of our beaches. How-

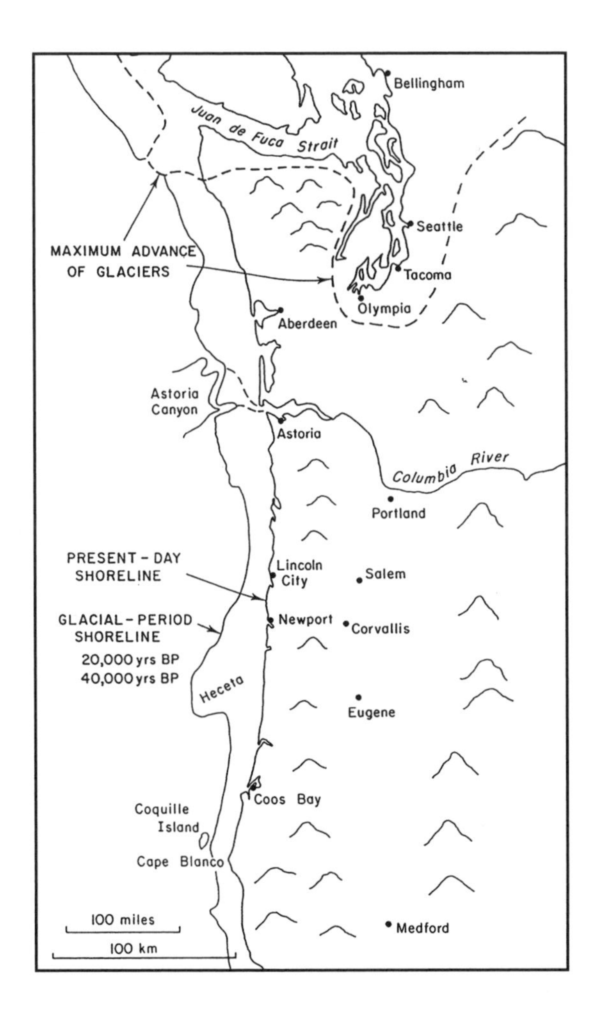

Figure 2.12 Approximate shoreline positions during low stands of sea level when glaciers reached their maximum development.

ever, the sands also contain small fractions of heavier minerals that are black, pink, various shades of green, and other colors. These grains are readily visible as specks in a handful of quartz-and-feldspar beach sand, although sometimes they are concentrated by the waves into black sand deposits on the beaches. These heavy particles can often be traced to specific sources (Clemens and Komar 1988a, 1988b). Most distinctive are the minerals derived from the ancient Klamath Mountains of southern Oregon and northern California (fig. 2.13). Klamath sands contain minerals such as glaucophane, staurolite, epidote, zircon, hornblende, hypersthene, and the distinctive pink garnet, which is often concentrated in pockets on the beach. In contrast, the rivers that drain the Coast Range transport sand that contains almost exclusively two minerals: dark green augite and a small amount of brown hornblende (fig. 2.13). Augite comes from volcanic rocks and is contributed to the rivers by erosion of the ancient seafloor rocks and intruded lavas high in the Coast Range. The Columbia River drains a vast

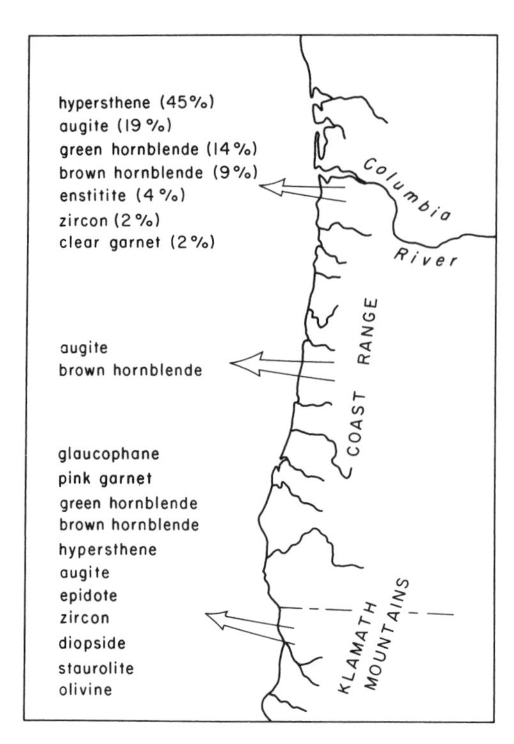

Figure 2.13 The principal suppliers of sand to the Northwest coast are the Columbia River, the Coast Range, and the Klamath Mountains. Each source supplies different heavy minerals to the beach and estuarine sands, and this makes it possible to assess their respective contributions. From Clemens and Komar 1988a.

area that contains many types of rocks, and this is reflected in the diversity of the heavy minerals in its sand (fig. 2.13).

The net northward movement of beach sand during lower sea levels (Scheidegger et al. 1971) is particularly indicated by the distribution of minerals derived from the Klamath Mountains. Sand from the rivers draining the Klamaths moved northward along the beaches under waves arriving from the southwest, and the minerals from that southerly source can be found in shelf sand nearly as far north as the Columbia River. As the Klamath-derived sand moved north, it combined with sand contributed to the beaches by rivers draining the Coast Range, so there is progressively more augite in the sand and a lower proportion of Klamath Mountain minerals. The Columbia River was a large source of sediment, but most of that sand moved to the north and so dominates the mineralogy of ancient beach sands found on the Washington continental shelf. Some Columbia River sand did move south along the Oregon beaches during times of low sea levels, mixing with the sand from the Klamath Mountains and Coast Range.

To summarize, the absence of headlands during lowered sea levels permitted sands derived from the Klamath Mountains, the Coast Range volcanic rocks, and the Columbia River to move north and south along the coast. Depending on their location along the former shoreline of the Northwest coast, beaches consisted of various proportions of mineral grains from these sources. Although some of the beach sand was left behind when the shoreline moved inland during the rapid rise in sea level and now rests on the continental shelf, much of it moved landward along with the migrating shoreline. The beaches would have been low in relief at that time, and storm waves would have washed over them, transporting sand from the ocean shores to the landward sides of the beaches, thereby producing the beach migration. Additional sand was contributed by rivers and by the coastal plain as it was eroded by the advancing sea.

About 5,000–7,000 years ago, the rate of sea level rise decreased as the water approached its present level (fig. 2.10). Also at about that time headlands segmented the Oregon coast into pocket beaches or littoral cells. At some stage several thousand years ago, the headlands extended into sufficiently deep water to hinder further alongshore transport of the beach sand (Clemens and Komar 1988a, 1988b). The pattern of along-coast mixing of sand from various sources established during lowered sea levels is still partly preserved, however, in the series of littoral cells that lie between the headlands. Minerals derived from the Klamath Mountains are still present in virtually all of the beaches along the Oregon coast, even though the sand can no longer pass around the many headlands that separate those beaches from the Klamaths.

At several locations on the Oregon coast there are distinct changes in beach sand mineralogies on opposite sides of headlands; that is, within adjacent but isolated littoral cells (Clemens and Komar 1988a, 1988b). The most dramatic change occurs at Tillamook Head south of Seaside, Oregon (fig. 2.14; Clemens and Komar 1988a, 1988b). The beach sand north of this headland is derived almost entirely from the Columbia River, and the abundant supply of sand has built out the shoreline significantly within historic times. The beach sand south of the headland contains abundant augite, indicating a Coast Range source from local rivers or cliff erosion. This beach sand also contains small amounts of Klamath Mountain minerals, the farthest north the relict pattern of along-coast mixing during lowered sea levels can be found preserved in the modern beaches. There is also some Columbia River sand in the beach south of Tillamook Head, but it got there by mixing southward with sand from the other sources during lowered sea levels and then migrating onshore. The Columbia River sand has been on the southern beach for thousands of years, but the beach sand north of the headland came from the Columbia River only within the last century or two. The different histories of the beach sands are substantiated by the degree of rounding of the individual grains. North of the headland the grains are fresh looking and angular, much like crushed glass, attesting to their recent arrival from the Columbia; the pounding surf has not had sufficient time to abrade and round the grains. South of the headland the grains are

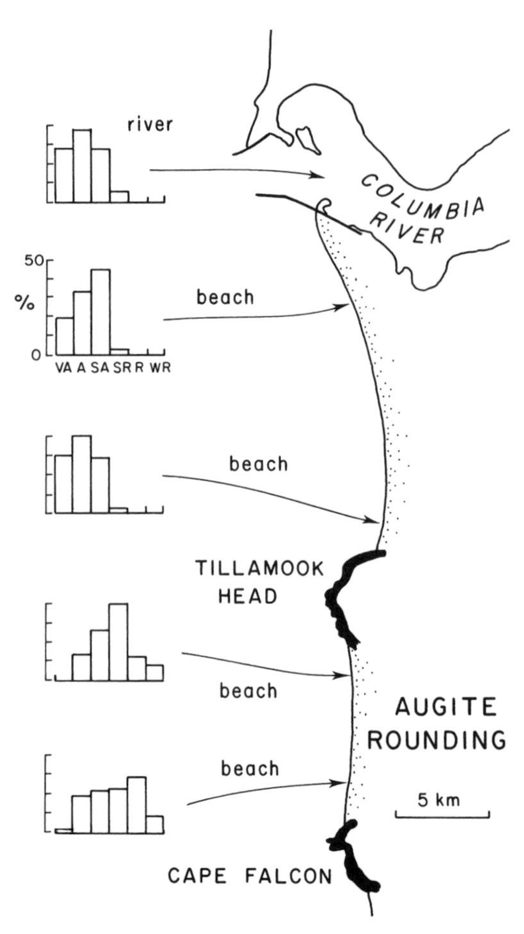

Figure 2.14 Changes in the degree of rounding of the beach sands on opposite sides of Tillamook Head. The grains to the north are more angular because they arrived from the Columbia River more recently. *Key:* VA = very angular, A = angular, SA = subangular, SR = subrounded, R = rounded, and WR = well rounded (1 kilometer ≈ 0.6 mile). From Clemens and Komar 1988.

much rounder, their sharp edges worn away during thousands of years of movement by wave swash on the beach (fig. 2.14).

The beach north of Tillamook Head is part of the Clatsop Plains (fig. 2.15), which formed after the sea reached its present level (Cooper 1958; Rankin 1983). The Clatsop Plains consists of beaches that have grown seaward together with a series of large dune ridges and associated interdune lows, many of which are occupied by elongated lakes. Backing the sandy plains is a wave-cut sea cliff that is now much degraded; it must have formed soon after the sea level rose but before the Columbia River could supply sufficient quantities of sand to develop a fronting beach. Dated peat deposits from within the plains indicate that the prehistoric shoreline fronting the eroded cliff formed about 3,500 years ago (Rankin 1983). The beaches built out rapidly until about 1,400 years ago, and then accumulated at a slower rate up to the present.

The Clatsop Plains is the only extensive stretch of shoreline on the Oregon coast that has naturally accumulated volumes of sand sufficient for the beach to advance seaward. This, of course, is due to the presence of the

Columbia River. The rest of the Oregon coast has more limited sources of sand, and those beaches have continued to erode. Most of the sand from the Columbia River moves north along the Washington coast, and as a result most of the Washington beaches have built seaward. The Long Beach Peninsula (fig. 2.15) is composed of Columbia River sand, and its northward extension is likely in response to a net longshore transport of sand in that direction (Ballard 1964; Komar and Li 1991). At the south end of the peninsula, where it attaches to the mainland, there is a small stretch of old sea cliff, which indicates that this sand spit built seaward just as did the Clatsop Plains. This seaward growth is even more evident farther north in the area of Grayland, where there is an old cliff along many miles of the coast separated from the present shore by a coastal plain 1–1.5 miles wide. For the most part, the Washington beaches are continuing to advance seaward by accumulating additional Columbia River sand, as indicated by comparisons of the present shoreline position with old coastal charts, and by annual beach profile surveys since 1951 (Phipps and Smith 1978; Schwartz et al. 1985).

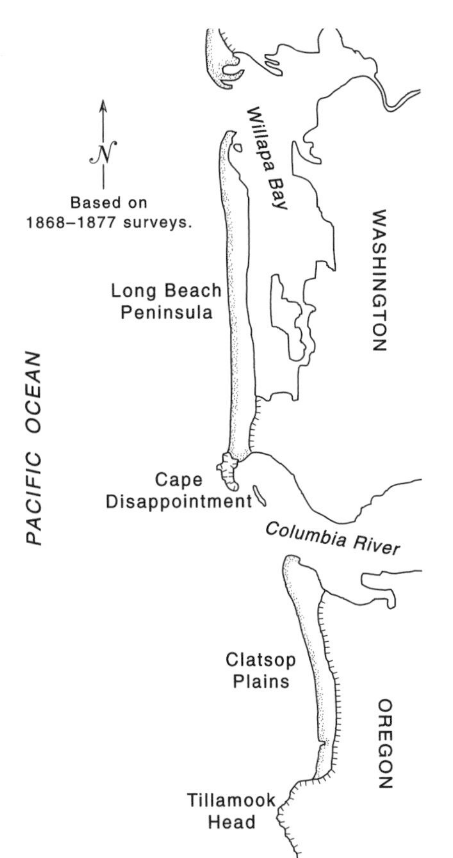

Figure 2.15 Chart of the Northwest coast from Tillamook Head, Oregon, north past the Long Beach Peninsula and Willapa Bay in Washington, based on 1868–77 surveys.

Formation of the Northwest Estuaries

The coastal rivers of the Northwest cut their valleys during the low sea levels that accompanied maximum glacial advances. When the water rose at the end of the ice ages, the valleys were drowned and developed into estuaries. Today these estuaries are important fisheries and harbors, and they serve as centers of community activity in many coastal towns. They also play a central role in sediment movements on the coast, which govern contributions of sand to the beaches.

Each estuary is connected to the ocean through an inlet, which carries out to sea fresh water that enters the estuary from the river (fig. 2.16). Also important to maintaining the inlet are tidal flows into and out of the estuary or bay.

Estuaries are zones where fresh water mixes with salt water. The fresh water is less dense and therefore tends to flow over the top of the seawater. At times, the fresh water may flow all the way through the estuary and into the ocean before finally mixing with the underlying seawater. When this occurs, the lens of denser salt water that lies below the fresh water exhibits a net flow from the ocean into the estuary. This type of flow, found in many Northwest estuaries, transports sediment from the ocean into the estuary and inhibits the movement of river sand from the estuary to the ocean beaches.

The restriction of sand movement through Northwest estuaries was first demonstrated in a study of sediments in Yaquina Bay (Kulm and Byrne

Figure 2.16 The Alsea Bay estuary on the mid-Oregon coast. A narrow inlet connects the bay to the sea.

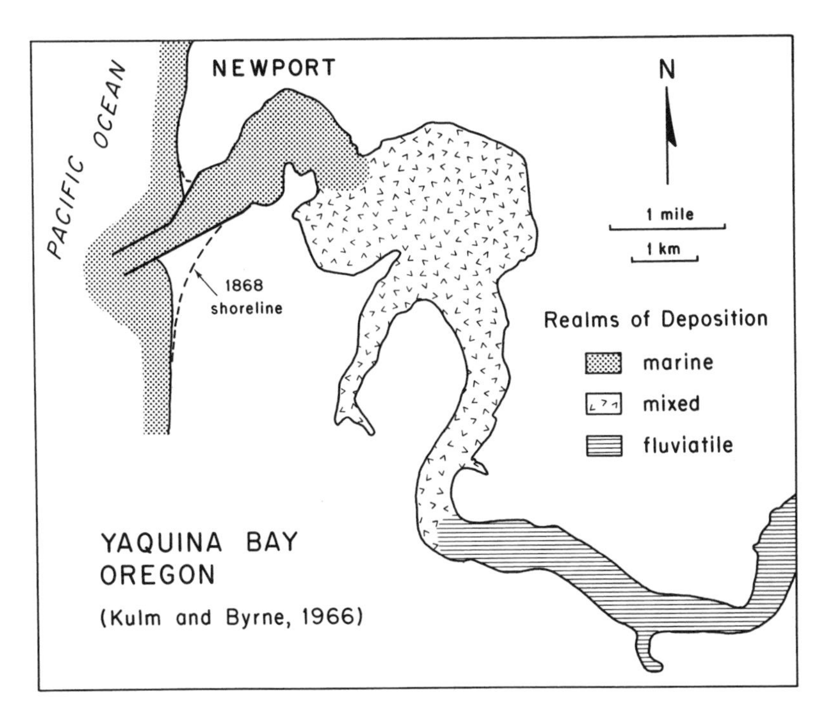

Figure 2.17 Sediment patterns within Yaquina Bay result from the mixing of marine sand carried into the estuary by tidal flows and fluviatile sand derived from the river. After Kulm and Byrne 1966.

1966). Like the other rivers draining the Coast Range, the Yaquina River carries sand that has augite as its principal heavy mineral. The beach sand outside the bay contains a large variety of minerals, including some derived from the Klamath Mountains. This difference in minerals makes it possible to trace the movement of the river and beach sands entering the estuary (fig. 2.17). River sand (fluviatile) forms 100 percent of the estuarine sediment only in the landward portion of the bay. Marine beach sand carried into the bay through the inlet dominates the estuarine sediments near the mouth. The remainder of the bay contains river and marine sands mixed in varying proportions. Yaquina Bay is slowly filling with sediment, both fluviatile sand from the landward direction and marine sands from the ocean side. Sediment cores indicate that Alsea Bay began to fill as soon as the estuary was formed after the last rise in sea level and is continuing to fill (Peterson et al. 1982, 1984b). In fact, this is the fate of most estuaries. Formed from drowned river valleys at the end of the ice ages, they are environments out of equilibrium. Most estuaries are eventually reduced to channels that transport all of the river sediments to the ocean. However, such a development takes thousands of years, so we should not view our estuaries as temporary features.

Although figure 2.17 indicates that little if any sand from the Yaquina River is presently reaching the ocean beach, this conclusion applies only to sand-sized grains. The fine clays remain in suspension in the water and are carried through the bay and into the ocean, evident in the brown plume that emanates from the inlet during river floods. Large estuaries separate most of the major coastal rivers of the Northwest from the Pacific Ocean and prevent them from contributing significant amounts of sand to modern beaches. This in part explains why many Oregon beaches have relatively small volumes of sand, and why their mineralogies still reflect the along-coast mixing of sand obtained from different sources during low stands of sea level rather than more recent contributions. The one clear exception is the Columbia River, which transports more than 100 times as much sand as the next largest river (the Umpqua), and on the order of 1,000 times as much sand as the other coastal rivers (Clemens and Komar 1988a).

Dune Fields of the Northwest Coast

Impressive accumulations of dune sands occur along the Northwest coast. In many areas the dunes are still being actively molded by winds; in other areas vegetation now covers formerly mobile dunes. It has been estimated

Figure 2.18 The Coos Bay dune sheet on the central Oregon coast. From the Oregon Highway Department.

Figure 2.19 The precipitation ridge at the landward edge of the Coos Bay dunes. The dune sand is slowly covering forests and blocking stream channels to form small ponds and lakes.

that sand dunes are present along 45 percent of the Oregon coast and 31 percent of the Washington coast (Cooper 1958). All of the coastal dune fields are within 2 miles of the ocean shore; most are immediately adjacent to sand beaches. It is clear that the dunes along the Northwest coast were formed by winds blowing sand inland from the beaches.

The Coos Bay dune sheet on the mid-Oregon coast is the largest coastal dune accumulation in the United States (fig. 2.18). It extends northward for nearly 150 miles from Coos Bay inlet to Heceta Head and is an impressive display of wind-blown landforms and processes.

There is evidence of three episodes of dune advance in the Coos Bay dune sheet, and also in other dune fields (Cooper 1958). The earliest is represented today by a strip of thoroughly vegetated dunes that in most places achieved the greatest landward advance. The second advance generally fell short of the first, and its present condition ranges from complete stabilization to still vigorous activity. The third episode is represented by the large areas of active dunes that until recently had open access to the ocean beaches that supplied them with sand. The landward edges of these dune fields are well defined by the presence of precipitation ridges (fig. 2.19), steep slip faces that slowly invade and bury forested areas. The precipitation ridge often blocks stream drainageways so that ponds and lakes develop along its base.

People have played a major role in altering the surface cover of the dunes along the Northwest coast. In some cases this has been deliberate, in others

Figure 2.20 Two views of the same area of the Clatsop Plains: *top*, active dunes in August 1937; *bottom*, July 1944, showing subsequent revegetation of the area using European beach grass. From Hanneson 1962.

Figure 2.21 Introduction of European beach grass in the Coos Bay dune field caused a foredune to build up at the back of the beach that cut off the inland movement of sand from the beach to the large dunes.

accidental. When Europeans first settled the Clatsop Plains south of the Columbia River in the nineteenth century, the extensive dune fields were covered by dense grasses (Hanneson 1962). The settlers' livestock ate the grasses, and overgrazing soon reactivated the dunes. By the 1930s, some 3,000 acres of sand had become mobile again (fig. 2.20), and blowing sand covered roads and homes. In 1934 the U.S. Soil Conservation Service took action to halt the advancing sand by first planting large areas with European beach grass (which livestock generally will not eat). Once the dune grass took hold, plantings of Scotch broom and shore pine followed. These activities have been successful in revegetating the Clatsop Plains and similarly disrupted areas.

This introduction of European beach grass had an unforeseen adverse consequence on the Coos Bay dune sheet. A hundred years ago these dunes existed as an unvegetated sand surface that extended from the ocean shore to the precipitation ridge at its landward edge. Sand was free to blow inland from the beach to supply material for the continued growth of the dunes. However, the European beach grass introduced during the 1930s in other areas of the coast quickly spread to the Coos Bay dunes and began to grow in the dunes immediately landward from the beach. These dune grasses captured sand blowing inland from the beach, resulting in the growth of high foredunes that have cut off the supply of sand to the large inland dunes. The impact was noted first in the area immediately landward of the foredunes (fig. 2.21), where the ground level was lowered to the water table.

This in turn permitted the growth of shrubs and other vegetation atypical of dune areas. The areal extent of the active dunes has decreased substantially, and there is concern regarding their long-term preservation.

Summary

The ocean shores of Washington and Oregon have seen dramatic changes over the eons. Compared with other coasts, the Northwest is geologically young—it came into being only within the last five million years. Erosion of the emergent land mass supplied sand and gravel to the coast, and these sediments accumulated to form beaches and fields of windblown dunes. Ice age glaciers sculpted the far north coast of Washington, cutting the Strait of Juan de Fuca. However, the main impacts of glacial advances and retreats on the coast were indirect. The accompanying rises and falls in sea level cut stairways of marine terraces and drowned river valleys to form estuaries. The Northwest coast today displays the cumulative evidence of the roles played by geologic, climatic, and oceanographic processes. Tectonic activity within the earth continues to cause its uplift and shakes the Northwest coast every few hundred years with a major earthquake.

3 The Dynamic Northwest Coast

The Northwest coast is one of the world's most dynamic environments. Severe storms strike during the winter, generating strong winds, driving rains, and huge waves that crash against the shore (fig. 3.1). Waves and ocean currents continuously reshape the shoreline. Sand is cut away in some spots and built up in others. Beaches give way, retreating inland and threatening homes, motels, and public parks. The reshaping of the shore can be understood only through a knowledge of ocean and beach processes—the waves, tides, and currents that interact with the land to modify shorelines. These change with the seasons of the year.

Seasons and the Coastal Climate

The difference between summer and winter in the Northwest is a difference of extremes. The summer months are dry with mild temperatures and winds. Vacationers can enjoy the sun and comfortable temperatures as they relax on the beaches or stroll through coastal towns. The weather begins to change in October, when temperatures fall and clouds and rain increase (fig. 3.2). Major storm systems generally begin to pass over the coast in November, bringing heavy rains and strong southwest winds. Only those who enjoy witnessing the awesome power of the winter surf breaking against the shore venture onto the beaches. The return to summer conditions is gradual. From January through May the temperatures progressively rise while the rainfall decreases (fig. 3.2). Intense storms seldom occur after March.

This extreme seasonality in climate creates parallel variations in ocean processes and is the strongest influence acting on the natural cycles that affect Northwest beaches. Winter storms generate much larger waves and stronger ocean currents than are seen in summer. The beaches erode under the onslaught of winter storms, then rebuild during the following summer.

Figure 3.1 Large waves breaking against the rocks of Cape Disappointment, Washington, along the north shore of the entrance to the Columbia River. Courtesy of the *Daily Astorian.*

Cliff erosion also increases during the winter, not only because cliffs are attacked by storm waves, but also because the rain washes away loose cliff materials and sometimes lubricates landslides. Examples of such cycles in coastal processes will be apparent throughout this chapter.

Potential buyers should be aware of the difference between the summer Northwest and the winter Northwest. New retirees arrive from the Midwest in summer to settle into the comfort of a beach home fronted by a wide beach and gentle surf, only to see the sand disappear during the next winter and the waves lapping at their doors.

Ocean Wave Generation

The energy carried by ocean waves parallels the seasonality of storm winds, because the strength of those winds is the primary determinant of wave heights. In general, the greater the velocity of the wind blowing over the ocean surface, the higher the resulting waves. Other factors are involved as well, of course. One is the duration of the storm: the longer the wind blows, the greater the energy transferred from the storm winds to the waves. Another factor is the fetch, the area or ocean expanse over which the storm winds are blowing. The importance of fetch is apparent when one compares wave generation on the ocean with that on an inland lake. The fetch on the

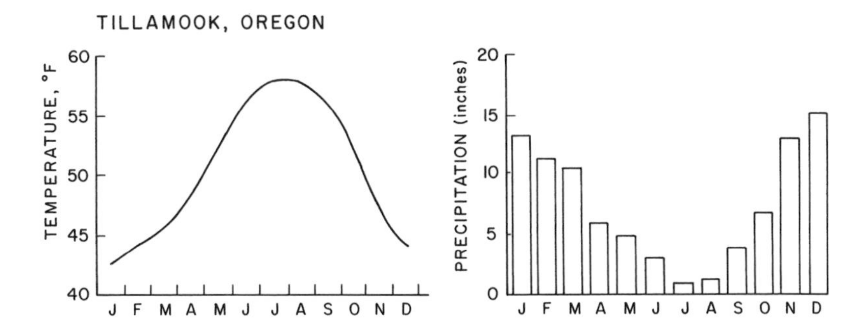

Figure 3.2 Seasonal variations in temperature and rainfall measured at Tillamook, Oregon.

lake can be no greater than its length, so the waves have time to acquire only a small amount of energy from the wind before they cross the lake and break on the far shore.

Wind-generated waves are energy transfer agents. They obtain energy from the wind, transfer it across the expanse of ocean, and finally deliver it to the coastal zone when they break on the shore (fig. 3.3). Therefore, the storm generating the high waves need not be in the immediate coastal zone. Waves reach the Northwest coast from storms all over the Pacific, even from the Southern Hemisphere near Antarctica. However, our largest waves are derived from winter storm systems that move down from the North Pacific and Gulf of Alaska.

Waves in the actual area of a storm are extremely irregular. Each wave changes rapidly in height, sometimes increasing but eventually decreasing until it disappears altogether. This complex pattern results because the storm generates many waves simultaneously; some are momentarily superimposed to produce higher waves, then a few seconds later they move apart

Figure 3.3 Wind energy is transferred to ocean waves, propagated in swell crossing the sea, and eventually delivered to the nearshore when the waves break.

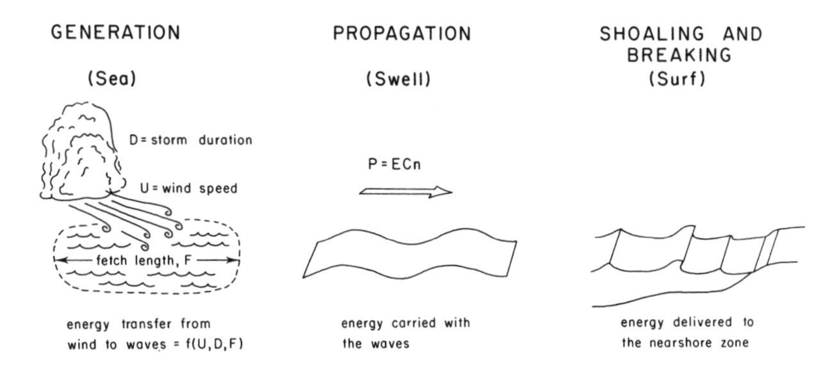

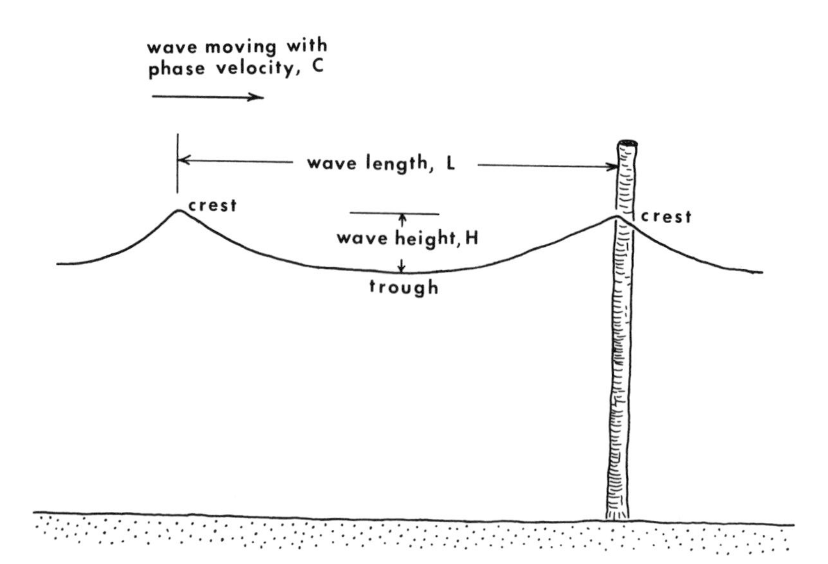

Figure 3.4 Simple ocean waves consist of a series of crests and troughs that are measured by the wave length, L, and height, H. The wave period is the time interval between the passage of successive waves, for example, past the piling shown. From Komar 1997.

and cancel one another. As the waves leave the storm area they begin to sort themselves and so become more regular. Waves distant from a storm are termed *swell* and are characterized by a series of crests and troughs (fig. 3.4). When waves are regular, like these, it is easier to measure their heights, lengths, and periods (the time it takes for successive crests to pass a fixed position such as a piling). Storm winds generate waves of many different heights and a wide range of periods. Long-period waves travel faster than short-period waves, and so move away from the storm area first and outdistance the short-period waves. This is a primary factor in converting the irregular waves found in the storm area into a regular swell. It also means that when waves generated by a distant storm reach the coast, the longest-period waves arrive first, followed by progressively shorter-period waves.

In general, the waves that break on the coast are a mixture of swell from distant storms and more irregular waves generated locally by coastal winds. In the absence of a local storm, which produces high sea conditions in the nearshore, it is the lighter coastal winds that produce the "chop" that is almost always present. Even when a strong chop due to local winds is present, however, the wave conditions along the coast still tend to be dominated by the regular swell generated in storms thousands of miles away.

If there are two distant storms, then two sets of swell waves may reach the shore at the same time. These two sets may be superimposed for a short time to produce larger breakers, only to cancel one another out a few minutes later, yielding lower waves. It is such combinations of swells from different storms that accounts for the common observation that the seventh wave is the largest; actually, the cycle can be every fifth wave to every tenth wave, depending on the difference in periods between the two sets of swells.

The variability of wave heights makes it difficult to characterize ocean waves. What does it mean, for example, when the Coast Guard reports 10-foot seas? One could report an average height of all the waves at the measurement site, but this is not generally done because the smallest are too insignificant to be included. Instead, the wave conditions are commonly reported in terms of the *significant wave height,* defined as the average of the highest one-third of the waves (Komar 1997). The significant wave height can be determined using sophisticated instruments, but it turns out that it also roughly corresponds to a visual estimate of a representative wave height. This is because an observer naturally tends to notice the larger waves and ignore the smallest. Of course, there will be many individual waves that are higher than the observed significant wave height, which is something of an average. The largest wave height during any 20-minute interval will be a factor of about 1.8 times the significant wave height (Komar 1997). Therefore, when the Coast Guard is reporting 10-foot waves, be prepared to face individual waves that are approximately 1.8 x 10 feet = 18 feet.

Ocean waves reaching the shores of the Northwest are measured daily with a unique system (fig. 3.5), a seismometer like those usually employed to measure earth tremors caused by earthquakes, but in this application tuned to sense the small ground movements produced by ocean waves as they reach and break on the shore (Komar et al. 1976b; Zopf et al. 1976; Creech 1981; Tillotson 1994; Tillotson and Komar 1997). A wave-measuring seismometer system at Oregon State University's Hatfield Marine Science Center in Newport is connected to a recorder to obtain a permanent record of the waves. This system has been in operation since November 1971 and is the longest continuous record of wave conditions on the West Coast in existence. Since the 1980s, the National Oceanic and Atmospheric

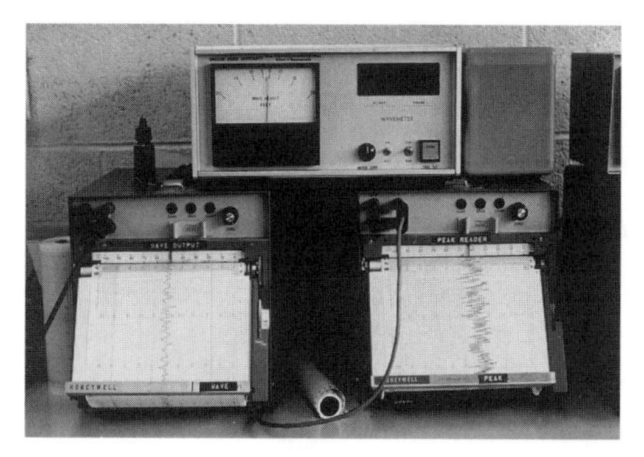

Figure 3.5 The seismometer at the Mark Hatfield Marine Science Center in Newport, Oregon, is tuned to measure ocean wave conditions.

Administration's (NOAA) National Data Buoy Program has maintained a series of wave-measuring buoys in deep water off the shores of the coastal states, including Washington and Oregon. Wave measurements derived from the seismometer, and more recently from the deepwater buoys, have been invaluable in research examining the causes of beach erosion in the Northwest.

It might come as a surprise that a seismometer in the Marine Science Center, which is 2 miles from the ocean, can provide records of ocean waves. But this seismometer differs from others in that it is tuned to amplify minor tremors, whether these are caused by earthquakes too small to be felt or generated by ocean waves. Even more impressive is the fact that the waves can be detected with a seismometer located at Oregon State University in Corvallis, 60 miles inland. When the surf is high on the coast, its effects show up as jiggles in the seismometer recordings. In order to use the record from the seismometer to measure ocean waves, it was first necessary to calibrate the system. This was accomplished by obtaining direct measurements of waves in the ocean at the same time their tremors were measured with the seismometer. The direct measurements of waves were collected with a pressure transducer, an instrument which rests on the ocean bottom and records pressures that are directly proportional to the heights

Figure 3.6 Daily variations in wave conditions on the Oregon coast measured with the seismometer, December 1972–January 1973 (1 meter ≈ 1.1 yards). From McKinney 1977.

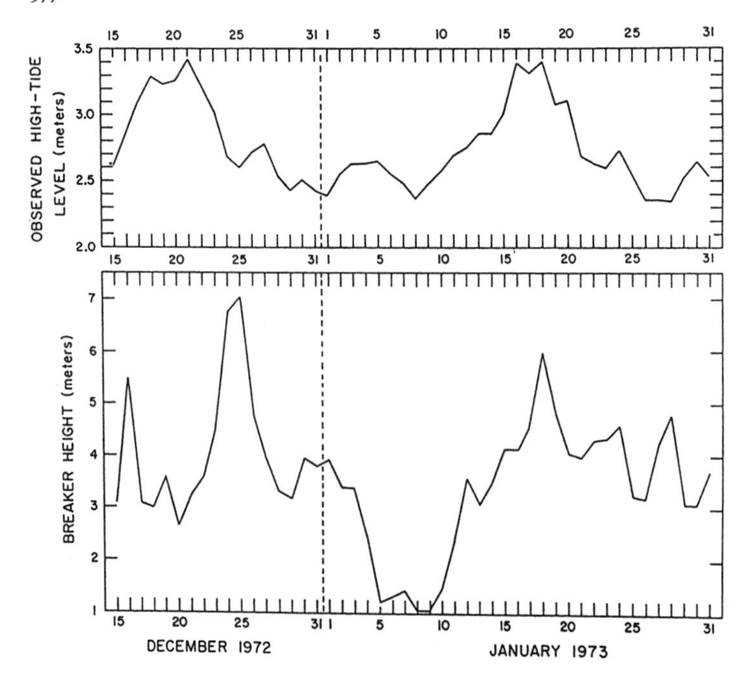

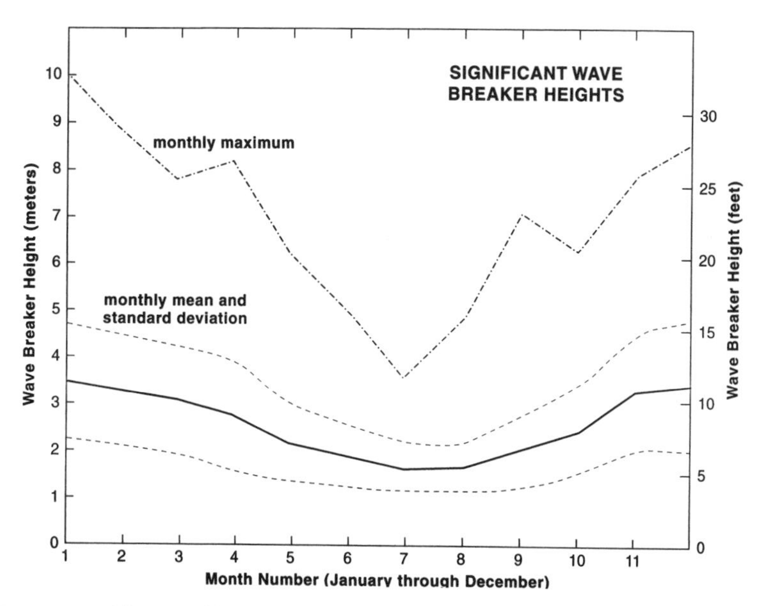

Figure 3.7 The monthly variations of wave breaker heights on Northwest beaches (1981–present), computed from wave heights measured in deep water by a NOAA buoy offshore from Newport. The solid line indicates the mean monthly heights (significant wave heights), the short-dashed lines are one standard deviation above and below the mean, and the long-dashed line is for the maximum monthly breaker heights measured since 1981 (1 meter ≈ 1.1 yards). After Tillotson and Komar 1997.

of the waves passing over the transducer. Now only the seismometer is needed to monitor daily ocean wave conditions, which it does safely from within the Marine Science Center, far from the destructive impacts of the waves.

Figure 3.6 shows daily wave measurements obtained from the seismometer from mid-December 1972 through mid-January 1973. Most apparent in this record are the storm waves that struck the coast at Christmas. The breaker heights reached about 25 feet—the height of a three-story building. But that figure represents the significant wave height (i.e., the average of the highest one-third of the waves), so there must have been individual waves with heights closer to 1.8 x 25 feet = 45 feet! As might be expected, there was considerable erosion along the coast during the December 1972 storm. The most severe erosion occurred at Siletz Spit on the mid-Oregon coast; we will examine that event in detail in chapter 6.

The seismometer in the Hatfield Marine Science Center and the NOAA buoys yield measurements of the wave conditions offshore in deep water. As the waves travel toward the shore they become taller and taller, achieving a maximum when they break on the sloping beach. Figure 3.7 is a graph of the yearly cycle of significant wave breaker heights experienced on

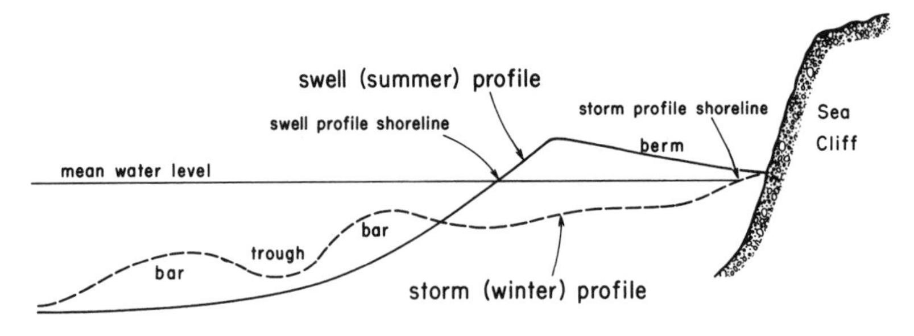

Figure 3.8 Seasonal changes in the beach profile in response to seasonal variations in wave energy. From Komar 1997.

Northwest beaches (Tillotson 1994; Tillotson and Komar 1997). The solid line gives the average breaker height for each month and shows that the breakers are on the order of 7 feet high during the summer, and nearly double that, about 13 feet, in the winter. Most breaker heights fall within the range delineated by the dashed lines on either side of the solid line. Of particular interest are the largest breaking waves generated by the most severe storms. The long-dashed line indicates the maximum monthly breaker heights and reveals the dramatic increase that occurs during the winter months. The highest storm wave conditions occur during December and January, when the maximum breaker height is on the order of 35 feet. Once again, this is the significant wave height; individual breaking waves could be on the order of 1.8 x 35 feet = 63 feet. Although this is an awesome height, even higher waves appear to have occurred on the coast in the not-too-distant past, but before the seismometer and buoys were in operation to measure the daily wave conditions. In the early 1960s, a wave-monitoring system located on an offshore oil-drilling platform measured a wave 95 feet high (Rogers 1966; Watts and Faulkner 1968). This is close to the 112-foot height of the largest wave ever reliably measured in the ocean, observed from a naval tanker traveling from Manila to San Diego in 1933 (Komar 1997). All of the measurements on the Northwest coast confirm that it has one of the highest wave energy levels in the world.

Beach Cycles on the Northwest Coast

Beaches respond directly to seasonal changes in wave conditions. The resulting cycle is similar on most coastlines (fig. 3.8). The beach is cut back during the winter months when high waves erode sand from shallow water and from the dry part of the beach. The eroded sand moves out to deeper water and accumulates in offshore bars located approximately in the zone where the waves first break as they reach the coast. The sand reverses its movement during the summer months, moving back onshore from the bars to widen the dry part of the beach. Although this cycle between two beach profile types occurs seasonally, the beach is really responding to the waves that strike the shore—high winter storm waves versus low regular swell waves. Sometimes low waves prevail during the winter and the dry beach may actually build out, although generally not to the extent of the summer beach. Similarly, beach erosion also occurs during summer storms.

The same cycle occurs on the beaches of the Northwest coast (Aguilar-Tunon and Komar 1978; Fox and Davis 1978), although factors other than wave energy also help to determine the size and shape of the beaches there. During the winter of 1976–77, my student N. A. Aguilar-Tunon and I measured profiles of two Oregon beaches: Devil's Punchbowl at Otter Rock and Gleneden Beach south of Lincoln City. We chose those beaches because of their contrasting sand sizes. The grain size is the primary factor governing the slope of a beach; the larger the grain size, the steeper the slope. Gravel beaches are the steepest, with slopes sometimes reaching 25–30 degrees. In contrast, the overall slope of a fine-sand beach may be only 1–2 degrees. Gleneden Beach has coarser-grained sand than Devil's Punchbowl (fig. 3.9) and is therefore steeper.

Figure 3.10 shows the month-by-month changes in the Gleneden Beach profiles and the rapid cutting back of the dry part of the beach as the winter storms became more frequent. The erosion began as early as October and continued through the spring. Sand did not return to the dry beach until April, May, and June.

The cycle of profiles at Devil's Punchbowl Beach was basically the same, at least in its timing, but the magnitude of the change was much smaller than at Gleneden Beach. Sand elevations at Gleneden changed by as much

Figure 3.9 Beach profiles from Gleneden Beach and Devil's Punchbowl Beach (Otter Rock), Oregon, on April 2, 1977, showing that the coarser-sand beach (Gleneden) is steeper. Vertical exaggeration is 10 times (1 meter ≈ 1.1 yards). From Aguilar-Tunon and Komar 1978.

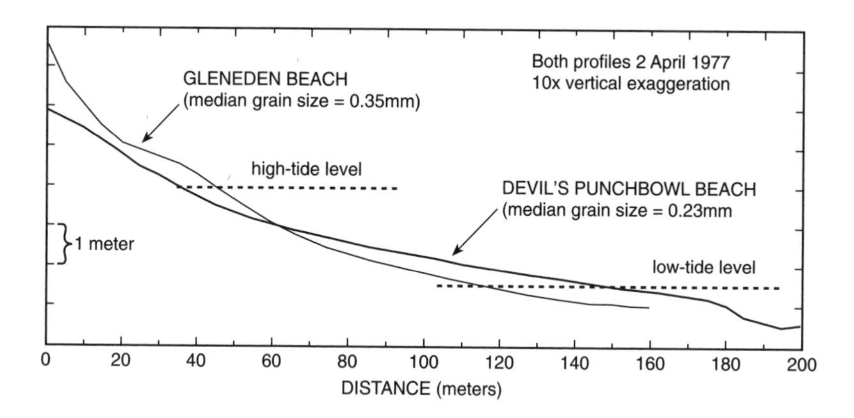

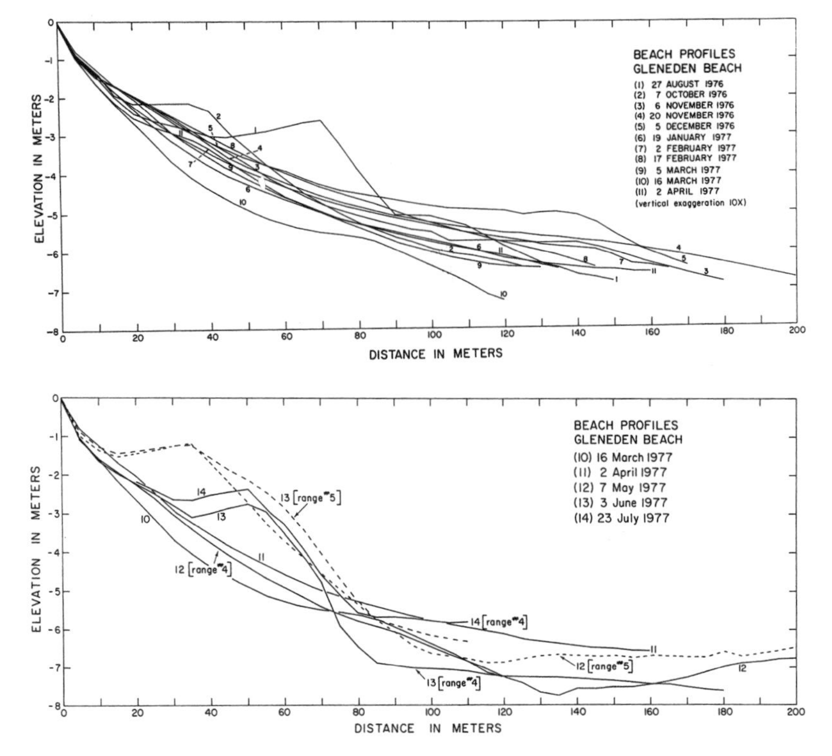

Figure 3.10 A series of beach profiles obtained at Gleneden Beach, Oregon, illustrating the seasonal variations on Northwest coast beaches (see fig. 3.9). There was considerable variation in the beach profiles along the length of the beach during May and June, as indicated by the different profile ranges included in the lower graph (1 meter ≈ 1.1 yards). From Aguilar-Tunon and Komar 1978.

as 8 feet (fig. 3.10), while the changes at Otter Rock amounted to less than 3 feet. In general, the coarser the beach sand (i.e., the larger the grains) the steeper the beach, and the more its profile changes in response to varying wave conditions. Coarser-grained beaches also respond to storms more rapidly than do finer-grained beaches—storm waves not only cut back the coarser beach to a greater degree, they also erode it at a much faster rate. The waves' energy is concentrated in a smaller area on a coarse-sand beach, which has a narrower surf zone than a fine-grained beach, whose low slope causes the waves to break farther offshore. Therefore, the same amount of wave energy strikes a substantially smaller area on a narrow coarse-sand beach; the wide surf zone of the fine-sand beach acts to dissipate the energy.

The greater response of coarser-grained beaches to storm waves is an important factor in coastal erosion. Waves that strike a coarse-sand beach are able to cut rapidly through the beach to reach the land and buildings behind it. This points to the general role of the beach as a buffer between the ocean waves and coastal properties. During the summer, when the dry beach is wide, the waves cannot reach beach properties, and erosion is not a problem. When the beach is cut back during the fall and early winter, however, it progressively loses its buffering ability and property erosion is more likely. Generally speaking, a storm that strikes the coast in October, when there may still be enough dry beach to serve as a buffer, will do much less damage to property than a storm that strikes from November through March, when the dry beach has disappeared and cannot serve as a buffer. In fact, however, the extent of the remnant beach is extremely variable along the coast, as is the parallel threat of property erosion. This variability in the width of the dry beach is the result of patterns of the nearshore currents that assist the waves in cutting back the beach.

We know much less about seasonal changes in the offshore sandbars, because it is difficult to measure beach profiles all the way out through the breaker zone along this high-energy coast. The best information we have concerning the offshore bars on Northwest beaches comes from profiles obtained during World War II as part of preparations for landings in the Pacific and on the Normandy beachhead in France. The profiles were acquired using an amphibious vehicle that could cross the beach and go out through the breakers (fig. 3.11). This was a dangerous operation, and on one occasion a breaker heaved the vehicle onto its side with its wheels pointing out

Figure 3.11 The surveying technique used to measure beach profiles in studies undertaken during World War II (employing an amphibious vehicle). From Bascom 1964.

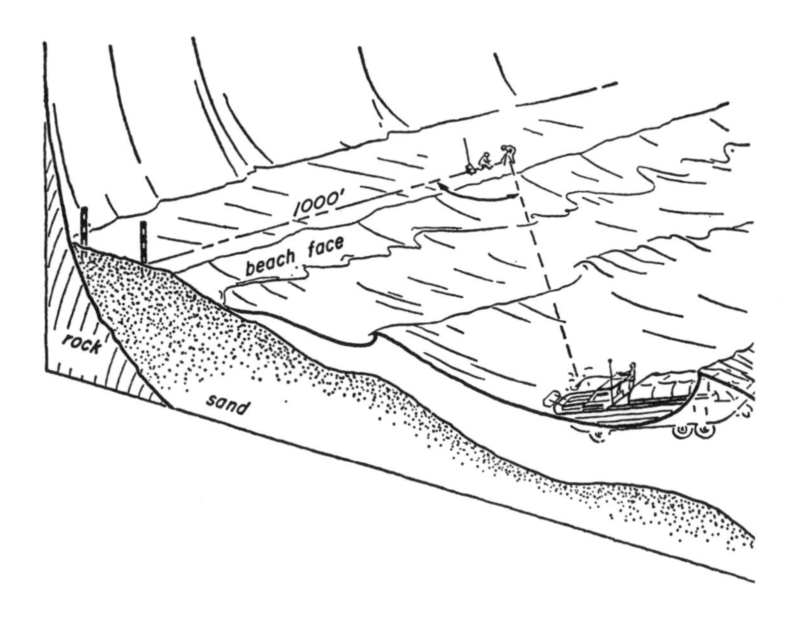

to sea (Bascom 1980). Fortunately, the next wave set it upright again without damage. Three amphibious vehicles were lost during the course of the study; two were lost in the surf and one rolled off a cliff and plunged 200 feet down into the sea. Amazingly, no one was seriously injured.

Profiles were obtained for nine beaches on the Washington coast and six on the Oregon coast (Komar 1978a). Figure 3.12 shows a profile from Solando, Washington, revealing a system of three offshore bars. (The profiles at other locations typically showed one to three bars.) When the profiling operation was repeated at Solando about three weeks later, the profile showed that the bars had grown and the troughs had deepened, probably in response to increased wave and current intensities. There was appreciable longshore variation in the size and even occurrence of offshore bars, the result of systems of nearshore currents and rip currents.

Nearshore Currents and the Movement of Beach Sand

Waves reaching the coast generate currents near the shore that are important to sand movements on the beach, and thus to erosion processes. These wave-generated currents are independent of ocean currents, which exist farther offshore and do not extend into the very shallow waters of the nearshore.

Most of the time, waves along the Northwest coast approach the beach with their crests nearly parallel with the shoreline. Under such circumstances the nearshore currents take the form of a cell circulation (fig. 3.13), with seaward-flowing rip currents being the most prominent feature. The rip currents are fed by longshore currents that flow roughly parallel with the shore, but only for short stretches. The currents of this cell circulation can move sand, and therefore can affect the beach's shape. The longshore cur-

Figure 3.12 Beach profiles from Solando, Washington, show three offshore bars, measured in 1946 with an amphibious vehicle, as shown in figure 3.12. From Komar 1978a.

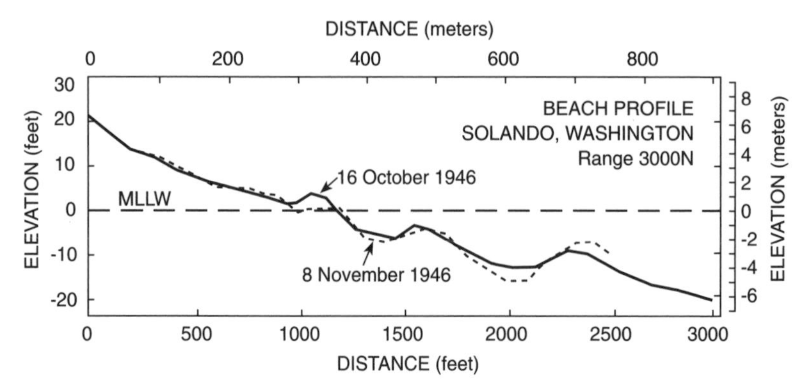

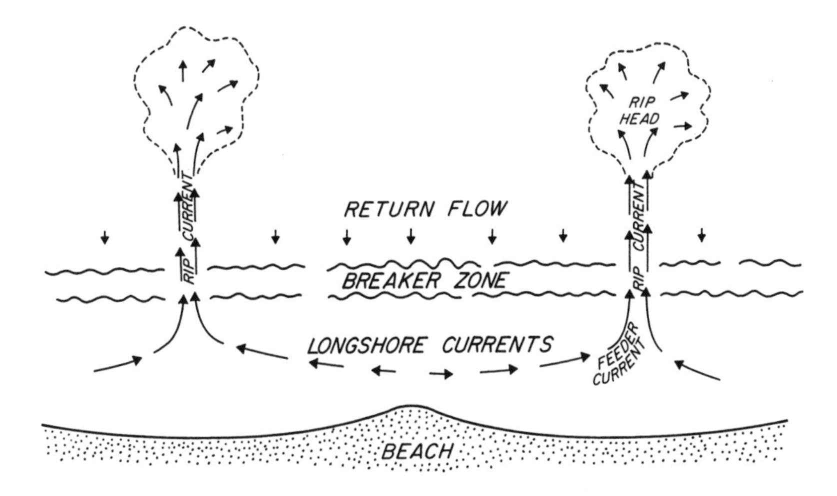

Figure 3.13 The nearshore cell circulation consists of seaward-flowing rip currents and longshore currents that feed water to the rips. From Komar 1997.

rents hollow out troughs into the beach, which generally increase in depth as a rip current is approached.

Rip currents can be very strong, cutting through the offshore bars to produce deeper water and a steeper but more uniformly sloping beach. They move sand offshore and erode crescent-shaped embayments into the beach. Aerial views of the beach typically show series of extremely irregular rip embayments of various sizes together with troughs cut by the longshore currents and rip currents (fig. 3.14). Rip current embayments sometimes extend across the entire width of the beach and begin to cut into foredunes and sea cliffs. Such rip embayments play a major role in erosion because they eliminate the buffering effect of the beach. Storm waves pass right through the deep water of the rip embayment and do not break and expend their energy until they reach the land behind the beach. Thus, a rip embayment may focus storm waves on a relatively small area (fig. 3.15). This type of erosion is commonly limited in extent to only 100 or 200 yards—the longshore span of a rip embayment that reaches the foredunes or sea cliff.

When waves approach the beach at an angle rather than head-on, they generate a current that flows parallel to the shore. Even then, however, seaward-flowing rip currents may be present. The longshore currents, together with the waves, transport sand along the beach in a type of movement known as *littoral drift*. This is more than a local rearrangement of the sand with an accompanying change in the shape of the beach. Littoral drift may displace sand hundreds of miles along a coast.

On the Northwest coast, the waves tend to arrive from the southwest during the winter and from the northwest during the summer (corre-

Figure 3.14 The beach at Nestucca Spit, Oregon, photographed during low tide, showing the troughs and embayments eroded by longshore currents and rip currents.

sponding to the prevailing wind directions). As a result, there is a seasonal reversal in the direction of littoral drift; north during the winter, south during the summer. The net littoral drift is the difference between these northward and southward sand movements.

Along most of the Oregon coast the net drift is essentially zero, at least if averaged over a number of years. This is demonstrated by the absence of continuous sand accumulation on one side of jetties or rocky headlands and erosion on the downdrift sides (Komar et al. 1976a). Patterns of sand accumulation and erosion on opposite sides of jetties (fig. 3.16, *top*) occur

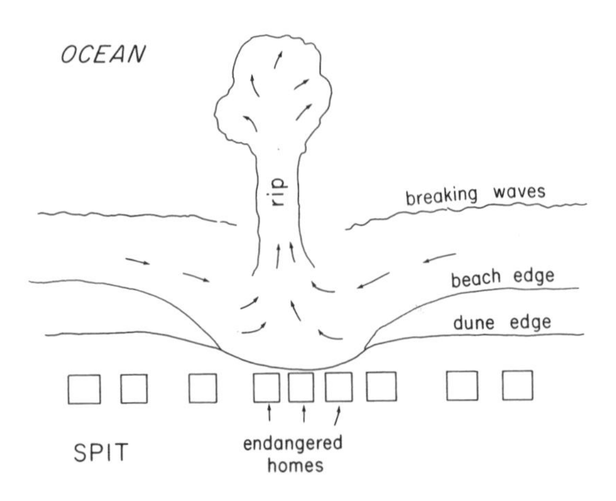

Figure 3.15 Rip currents erode embayments into the dry beach and locally threaten properties backing the beach.

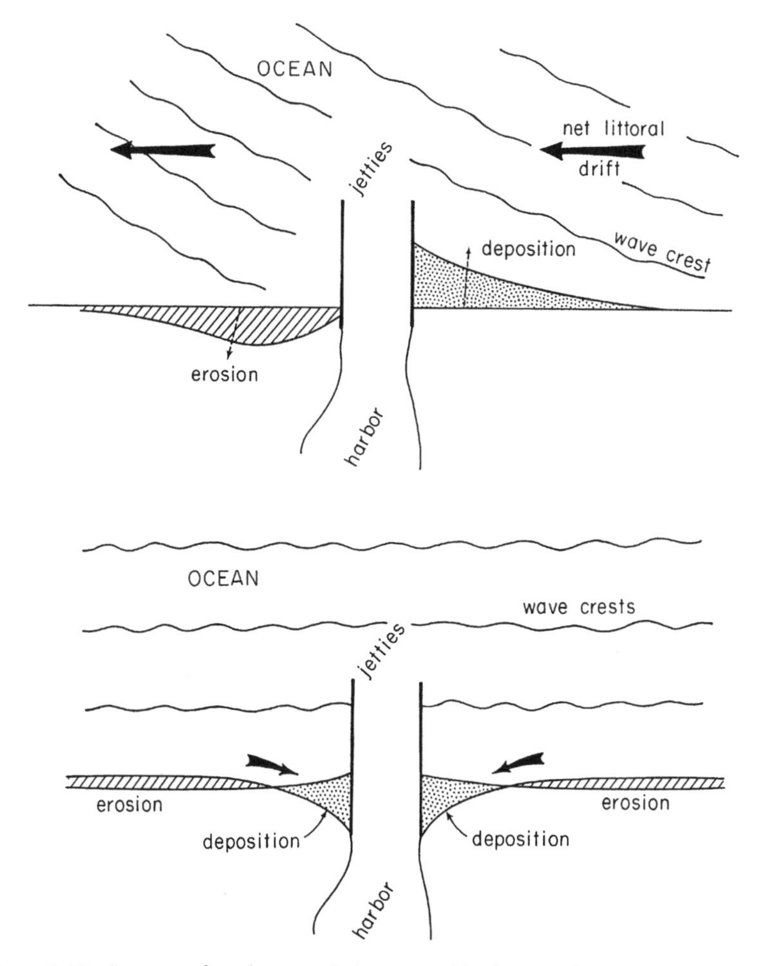

Figure 3.16 Patterns of sand accumulation around jetties: *top,* the jetties are blocking net littoral drift, and sand accumulates in the updrift direction and erodes downdrift of the jetty; *bottom,* there is no net littoral drift for the jetties to block. Jetties on the Oregon coast exhibit the latter condition.

on coasts where there is a net littoral drift; for example, along the shores of southern California and most of the east coast of the United States. Jetties on those coasts act like dams to the longshore movement of beach sand, accumulating sand on their updrift sides and causing erosion downdrift. This often causes major losses of beach and property. Jetties built on the Oregon coast, which has no net littoral drift, tend to accumulate sand on both sides (fig. 3.16, *bottom*). The accumulated sand comes from small-scale erosion of beaches more distant from the jetties, so that an overall symmetrical pattern emerges. The shoreline soon reaches a new equilibrium, so the erosion does not continue for very long. In sum, the Northwest coast is less likely than other coasts to experience major erosion and property losses due to the construction of jetties. One severe erosion problem *did* occur on the Oregon coast in direct response to jetty construction, however, and it led to the destruction of the town of Bayocean in the early half of this century (see chapter 5).

The reason for the zero net littoral drift of sand along the Oregon coast is that the beaches are contained within pockets between rocky headlands that are large enough and extend into sufficiently deep water to prevent beach sand from passing around them (see chapter 2). The sand in each pocket beach is isolated within its cell. It may move north and south within its pocket in response to seasonal wind and wave directions, but the long-term net movement is zero.

The one beach on the Oregon coast that does not fit the pattern of a zero-drift pocket is the Clatsop Plains, the shoreline that extends south from the Columbia River past Seaside to Tillamook Head. The plains were formed by the accumulation of sand derived from the Columbia River, a small part of which moves south until it is blocked by Tillamook Head. The bulk of the Columbia River sand moves northward along the coast of Washington. Most of it stays on the beaches just north of the Columbia River, but some continues farther north, in decreasing amounts, until beyond Copalis Head net erosion prevails.

On many coasts, sand spits grow in the direction of the net littoral drift. The Long Beach Peninsula, which extends northward from the Columbia River, likely reflects the net sand movement along the Washington coast. It is unclear whether this northward growth has continued within historic times, however, because there have been many cycles of growth and erosion at the tip of the peninsula. Some of the shifts at the end of the spit have been caused by migrations of the entrance to Willapa Bay, which has produced major erosion problems at Cape Shoalwater just north of the inlet (see chapter 6).

There are a number of sand spits along the northern half of the Oregon coast; some point north, others point south. Those spits are located within the beach cells where zero net littoral drift prevails, and their directions do not provide testimony of net longshore sand movements.

Tides along the Northwest Coast

The oceanographer Albert Defant described tides as the "heartbeat of the ocean" (Defant 1958), an appropriate characterization of these rhythmic rises and falls, predictable and unexciting. Tides along the Northwest coast are moderate, with a maximum range of about 13 feet and an average range

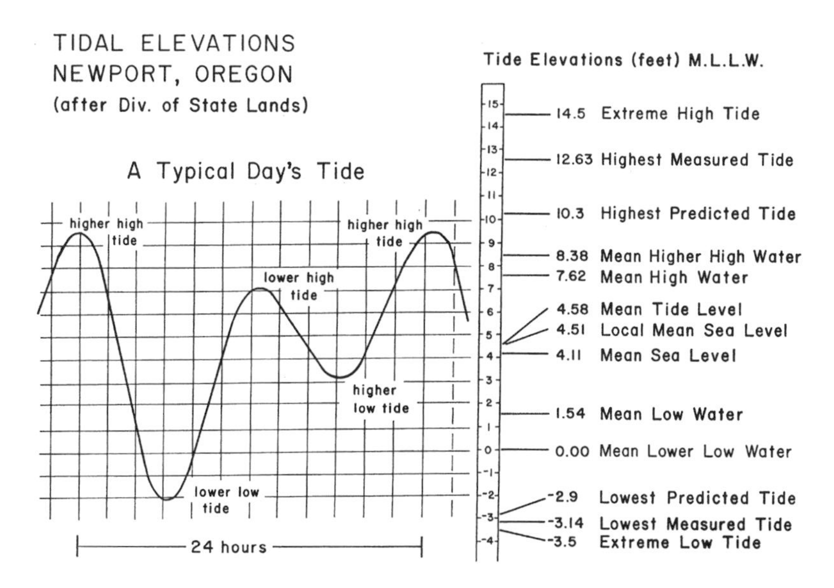

Figure 3.17 Daily tidal elevations measured in Yaquina Bay on the mid-Oregon coast are typical of tides along the Northwest coast. From Hamilton 1973.

of 6 feet. There are two highs and two lows each day, with successive highs (or lows) usually having markedly different levels. Tidal elevations are given in reference to the mean of the lower low water levels (abbreviated MLLW). Accordingly, most tidal elevations are positive numbers; only the extreme lower lows have negative values. In Yaquina Bay, Newport (fig. 3.17), the highest predicted tide is 10.3 feet MLLW and the lowest predicted low is -2.9 feet MLLW, giving a total range of 13.2 feet.

When the moon, sun, and earth form a straight line in space (termed "syzygy"), their gravitational forces combine to produce the highest monthly tidal range, the *spring tide.* This alignment occurs at full moons and new moons, and spring tides thus occur each month. Once a year the moon makes its closest approach to the earth (termed "perigee"), and its gravitational attraction on the ocean reaches a maximum. The resulting *perigean spring tide* is accordingly the largest range of the entire year—this is the 13.2-foot maximum range for Yaquina Bay. When the earth, moon, and sun are not aligned, the net gravitational forces are reduced and the tidal ranges are lower. The lowest tidal range occurs when the moon and sun are at right angles with respect to the earth, so their tide-producing forces are in direct opposition and tend to cancel; this lowest tidal range of the month is called the *neap tide.*

Tides are an important factor in coastal erosion because they govern the hour-by-hour level of the sea and hence the position of the shoreline and the zone where the ocean waves expend their energy. Spring tides, and especially perigean spring tides, may bring water levels high up on sea cliffs and foredunes so that waves reach and attack coastal properties. High spring tides contributed to the breaching of Nestucca Spit, Oregon, in 1978 (see chapter 6) and are a significant factor in sea cliff erosion (see chapter 8). Tides also affect the currents in bays and estuaries. In bays such as Netarts and Willapa, which receive little freshwater flow from rivers, the tidal currents are the sole factor keeping the inlets open to the sea.

Water Level Fluctuations

The measured water levels are usually different from those predicted by tide tables. Tidal computations are made using ideal conditions and consider only the gravitational forces of the moon and sun together with the influence of the overall shape of the ocean basin. The "errors" in the tide table predictions result from other processes that act on water elevations, such as winds, atmospheric pressures, ocean currents, and water temperatures.

On many coasts, the most significant water level changes are brought about by meteorological factors such as strong winds and changes in atmospheric pressure. When the wind blows toward the shore, water can pile up against the coast and reach levels above the predicted tides; offshore winds lower water levels. The stronger the wind and the longer its duration, the greater its effect on the water level. If the winds are part of a storm system, reduced atmospheric pressure is another contributing factor because storms are associated with low-pressure centers. The local reduction of atmospheric pressure causes a "humping" of the water on the ocean surface, which acts as an inverse barometer. Together the winds and reduced atmospheric pressures of a storm can create a *storm surge*. As the name implies, this is a surge of water over low-lying areas of the coast, sometimes reaching levels several feet above normal. Storm surge is most dramatic during hurricanes and cyclones, and has been the cause of extreme destruction and loss of life along the east and Gulf coasts of the United States. Fortunately, storm surge has not been a significant problem in the Northwest. The rise in water level beyond that expected from tides amounts to only 6–12 inches on our shores (McKinney 1977). Although small, such increases in water levels may nevertheless intensify the processes of coastal erosion. Many of our beaches have very low slopes, and even a small vertical rise in the water level due to a storm surge can produce a significant landward shift of the shoreline—in some cases 50–100 feet—placing the water closer to beach properties and allowing ocean waves and currents to act more directly against the land.

Water levels along the Northwest coast are consistently higher during the winter than in the summer (fig. 3.18). The average water levels may be anywhere from 4 to 15 inches higher than expected from the tides alone, but

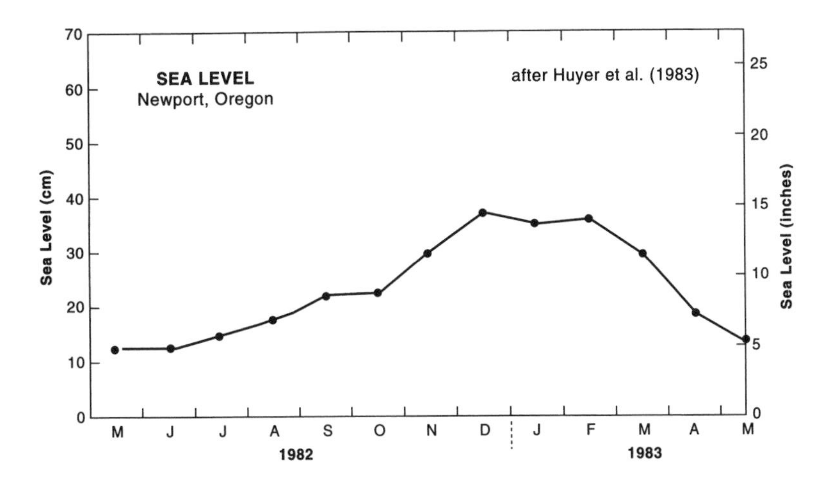

Figure 3.18 Monthly variations in mean sea levels measured by the tide gauge at Newport, Oregon, illustrate the annual cycle, with higher water levels during the winter (10 centimeters ≈ 4 inches). After Huyer et al. 1983.

they are always greatest during the winter. The data shown in figure 3.18 were obtained from the tide gauge in Yaquina Bay on the mid-Oregon coast, determined by averaging the actual measured water levels. This averaging has the effect of removing the daily tidal fluctuations, so the averages given represent deviations from the worldwide sea level as well as from the tide level in Yaquina Bay during any given hour.

A number of processes combine to produce the higher winter water levels, including storm surge (because onshore winds tend to be strongest in the winter) and seasonal variations in ocean currents and water temperatures. During the winter, the ocean currents off the Northwest coast flow primarily northward in response to the prevailing winds of that season. Although the dominant current direction is parallel with the coast, there is a component directed toward the shore, caused by the earth's rotation. When a wind produces a current in the Northern Hemisphere, the flowing water tends to rotate slightly to the right of the wind's direction (in the Southern Hemisphere the rotation is to the left). This onshore component of the northward current during the winter raises water levels along the shores of Oregon and Washington. In the summer the currents flow toward the south and the rotation is directed offshore, lowering the sea level along the coast.

Another factor that contributes to the lower summer water levels is coastal upwelling produced by the offshore component of the ocean currents, which moves the near-surface warm water away from the land. The warm water is replaced by cold water that moves up from the depths, with several consequences. When the cold water meets the warm air, fog is produced. The cold water contributes to the lowering of ocean surface levels along the shore during the summer (fig. 3.18) because it is denser than the warmer water that covers most of the ocean.

The ocean waves themselves raise water levels in the nearshore, an increase that is called *wave setup*. The energy of the waves approaching the beach is partly expended in causing the mean water level to slope upward toward the shore. The effects of the wave setup are maximum at the shoreline itself and are reduced to approximately zero at the breaker zone. The higher the breakers on the beach, the higher the setup above the horizontal still-water level (the level that the sea would have if there were no waves). The maximum setup at the shoreline is about 0.17 times the height of the offshore ocean waves (Guza and Thornton 1981; Holman and Sallenger 1985; Komar and Holman 1986). Accordingly, 10-foot waves will produce a setup of 1.7 feet, or 20 inches, at the shoreline. If a storm should increase the wave height to 20 feet, the setup elevation would be raised to 40 inches. A rise in setup can cause a significant landward migration of the shoreline. On a 1-in-50 beach slope, a 20-inch rise during a storm would move the mean shoreline landward by more than 80 feet.

Wave Run-up and Sneaker Waves

You might also expect storms to cause a considerable increase in the energy of the waves at the point where they swash up and down the beachface. But here nature has a surprise, at least on beaches with a low slope. Ocean waves directly generated by winds have periods less than 20 seconds. Therefore, the interval between successive breaking waves will be 20 seconds or less, as will the swash of the wave surf after the bores have crossed the beach and run up on the shore. But if we actually measure the swash run-up on a beach with a low slope under such conditions, we find that an increase in breaker heights does not produce the expected increase in swash energy and height at the shoreline (Guza and Thornton 1982; Holman and Sallenger 1985; Komar and Holman 1986). Why not? A wave breaks as it approaches the shoreline at a point where the depth of the water is approximately equal to the wave's height; that is, a 10-foot wave will break in approximately 10 feet of water. Thus, if the offshore waves double in height during a storm, they break in water that is twice as deep, and roughly twice as far offshore. The surf bores now have farther to travel before reaching the shoreline, losing energy as they go. Thus, it seems clear that a doubling of the breaker heights will not double the swash run-up on the shore.

At this point nature adds a further complication, however, one that is not fully understood. Even though the run-up of wave motions with periods less than 20 seconds does not increase significantly when breaker heights increase, there is a considerable increase in run-up associated with water level motions having periods greater than 20 seconds. As the ocean waves approach the coast and break, all of their energy is contained within wave oscillations that have periods less than 20 seconds. But somehow that energy is transferred to longer-period motions within the surf zone; it is the mechanism of this energy transfer that is poorly understood. The swash run-up related to these longer-period cycles does depend directly on the offshore breaker heights and energy, so it is clear that normal ocean waves are at least indirectly the source of this energy.

The long-period cycles are observed as horizontal movements of the mean shoreline, especially during storms: the shoreline slowly migrates up the beach, reaches a maximum landward position, and then slowly shifts seaward, only to return again. The entire cycle may take 2–5 minutes. Superimposed on that cycle is the swash of the incoming wave bores that have the shorter periods of regular ocean waves (i.e., less than 20 seconds). Accordingly, the zone of wave swash and run-up migrates slowly back and forth across the beach.

This is the process that creates "sneaker waves," so called because they sneak up and suddenly drench a person walking along the beach. You may be on what you think is the dry part of the beach, well above the action of the wave swash. But the long-period motions cause the water's edge to shift slowly landward, and if there is suddenly a larger-than-average wave bore, you may abruptly be overtaken by the combined run-up—you have experienced a sneaker wave. The consequences can be serious. The water may suddenly be up to your knees or deeper, accompanied by strong currents and wave swash. The danger to children is obvious. Some people try to escape the water by jumping atop a drift log, only to have the waves roll the log over them. Lives have been lost in this way.

Tsunami: The Extreme Coastal Hazard

There is one type of destructive wave that is not generated by winds: the tsunami, often incorrectly called "tidal wave" (tsunamis have nothing to do with tides). Tsunamis are produced by a displacement of the seafloor by an earthquake, explosive volcanic eruption, or submarine landslide. The sudden upward or downward movement of a portion of the seabed momentarily raises or lowers the overlying water surface. That disruption then travels outward from its site of origin as a series of waves moving at velocities of 350–500 miles per hour. In the deep ocean these waves are small, typically less than 3 feet, too small to be noticed as they pass beneath a ship. But tsunamis have very long periods; the time between successive waves in a series is 10–20 minutes. One effect of this long period is that tsunamis "feel" the bottom as soon as they reach the continental shelf. The shallower water slows their rate of movement, which causes their heights to progressively increase. As the tsunami waves cross the wide continental shelf, they get higher and higher, until at the shore itself they may achieve heights up to 50 feet. Their breaking and swash on a sloping beach can carry water far beyond the normal reach of the ocean.

The first sign of an impending tsunami is often an unusual and sudden lowering of the water level along the shore. Basically, this low water is a trough between the crests of the tsunami waves. Each wave is separated

Figure 3.19 Heights of tsunami waves measured within bays and estuaries along the Northwest coast on March 28, 1964. The solid line shows the approximate location and time of the tsunami wave front; the dashed lines show the approximate locations and times of the spring tide crest. From Wilson and Torum 1988.

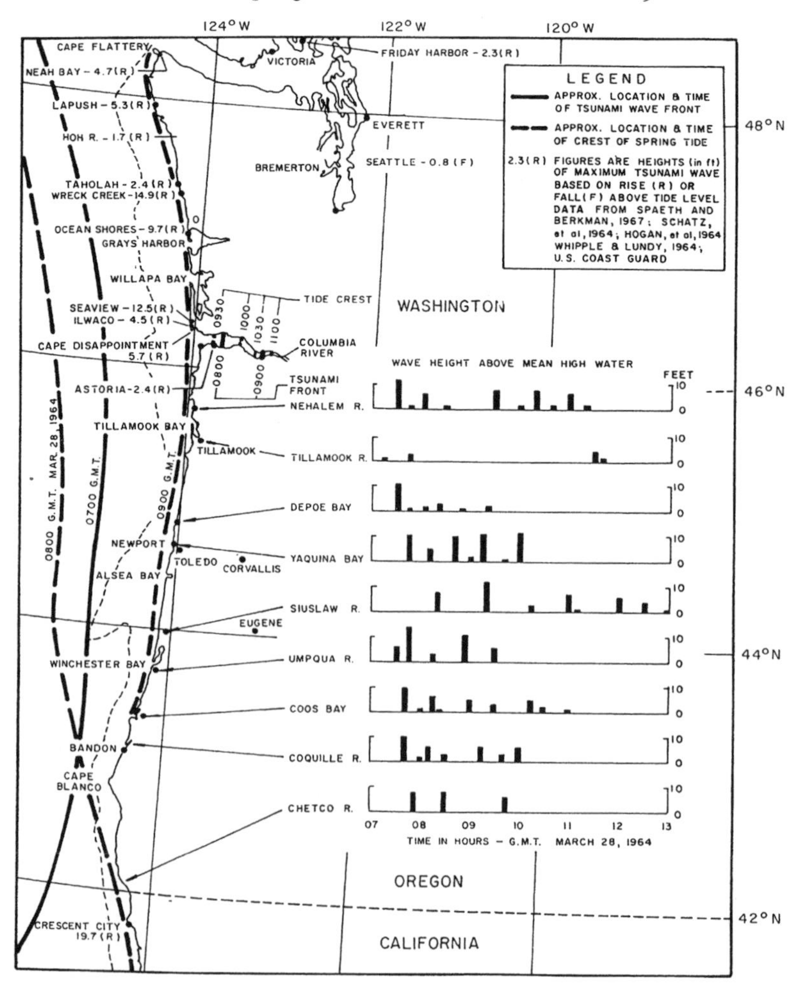

from the one preceding it by such a low. Just as it would be extremely unwise to stay on the beach when such a sudden lowering occurs, so it is also not advisable to rush out onto the beach after the first wave to see the resulting damage.

Most of the significant tsunamis that reach the Northwest coast are caused by earthquakes in and around Alaska. Two struck our shores during the 1960s—on March 28, 1964, and May 16, 1968 (Schatz et al. 1964; Schatz 1965; Wilson and Torum 1968). Figure 3.19 shows the heights of the tsunamis during the 1964 event as measured on tide gauges in several bays; unfortunately, no measurements are available for the open coast, where the wave heights undoubtedly were much greater. Each location experienced a series of waves, and the first was not necessarily the largest (fig. 3.19). Maximum wave heights in the bays and estuaries reached approximately 10 feet, an extremely high level for those sheltered environments.

The 1964 tsunamis damaged bridges and dwellings along the shores of the Necanicum and Neawanna Rivers where they flow through Seaside, Oregon; the cost of the damage was assessed at $276,000 (Wilson and Torum 1968). Other hard-hit areas were Cannon Beach ($230,000), Waldport on Alsea Bay ($160,000), Florence ($50,000), and Coos Bay ($20,000). Fortunately, there were few reports of destruction along the open-ocean shores of Oregon and Washington during the 1964 and 1968 tsunamis. Four children were drowned as their family slept on the beach at Beverly Beach State Park in Oregon when the 1964 tsunami struck. Greater damage occurred at Crescent City in northern California. During the 1964 tsunami, one wave washed about 500 yards inland, destroying 29 blocks of the business district and causing some $29 million in damage. However great, those impacts are substantially smaller than those that have been experienced in Japan, Hawaii, and Alaska, which are periodically devastated by tsunamis.

Hanging over the Northwest coast is the threat of a truly catastrophic tsunami that could be generated by a local subduction earthquake. As discussed in chapter 2, there is firm evidence that a major subduction earthquake occurs immediately offshore every few hundred years. These must generate tsunamis of awesome proportions. The main evidence that such events have transpired lies in the layers of sand carried inland from the beaches. Dr. Curt Peterson of Portland State University has examined such layers at several sites. Figure 3.20 shows his map tracing the surge of the tsunami that occurred about 300 years ago (in the year 1700) as it moved up into Yaquina Bay (Peterson and Priest 1995). The main path of the giant waves was along the length of the bay and up the river; sand layers attributable to those waves have been found as far upriver as Grassy Point, 8 miles from the mouth of the bay. The flows also surged up side channels as the tsunami passed through the bay (fig. 3.20).

Even elevated areas are not safe from inundation by tsunami waves asso-

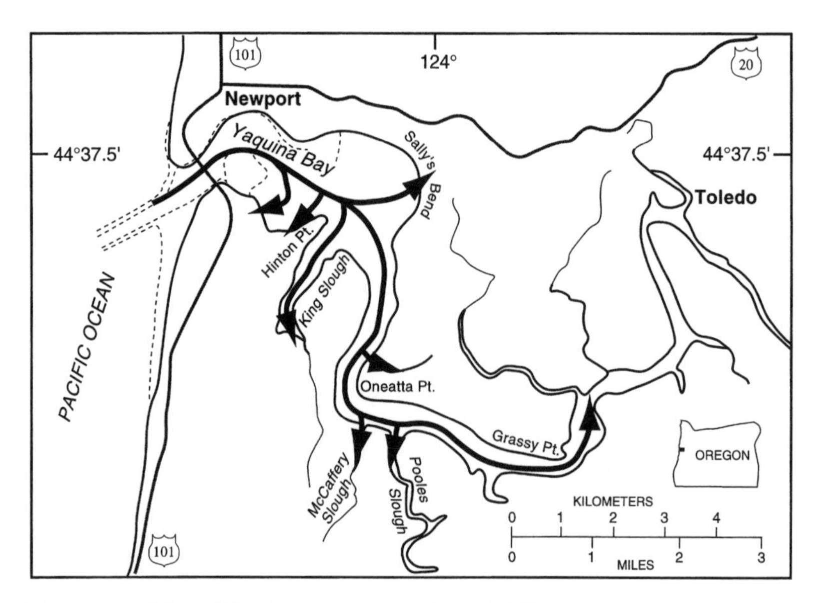

Figure 3.20 Map of Yaquina Bay showing the path followed by the tsunami surge generated by the subduction earthquake in 1700. Evidence of this penetration comes from layers of beach sand that covered marshes. From Peterson and Priest 1995.

ciated with subduction earthquakes. Cores drilled by Dr. Peterson throughout the city of Cannon Beach reveal the existence of a sand layer that was laid down by the 1700 tsunami. The layer extends inland to at least Highway 101 east of the city, evidence that the run-up of the tsunami and its destructive impacts can reach more than a mile inland even when passing over elevated areas.

An early warning system assesses whether or not earthquakes in such distant areas as Alaska and Japan have the potential to generate tsunamis. The system's predictions are not absolute, however, and there is usually a wait-and-see period before it can be established that destructive waves have actually formed. The system provides adequate warning time for people to evacuate beaches and other low-lying areas, but it also has the opposite effect: people flock to the coast to see the giant waves. After a recent warning the roads were clogged by thousands of cars carrying people from the valley to the Oregon coast. Fortunately, it turned out to be a false alarm. Such false alarms may create a "cry wolf" syndrome and make people less responsive to warnings and less heedful of the danger, even in areas that have been devastated by tsunamis in the past.

People can be evacuated from areas threatened by a tsunami, but their homes and possessions remain in the zones of potential destruction. Since the last tsunami in 1968, there has been a great deal of construction on the

Northwest coast. Many more homes now line ocean beaches and the shores of bays and estuaries. It can be expected that future tsunamis will cause considerably more property losses than in 1964 and 1968.

Many communities have developed local warning systems and evacuation routes to prepare for future tsunamis associated with subduction earthquakes. The earthquake itself will, of course, provide the first warning. People who live in hazardous zones must be made aware that the earthquake will likely be followed by a tsunami. The first wave can be expected to arrive 10 to 20 minutes after the quake.

Summary

The Northwest coast is shaped by powerful ocean and beach processes. Winter storms generate huge waves that combine with rip currents to erode beaches and attack coastal properties. The wave swash at the shore is the biting edge of the sea, the area where the conflict between the ocean and land reaches a maximum. During the winter the zone of wave swash moves landward due to a number of processes that combine to raise water levels in the coastal zone. This inland movement can bring erosion processes directly up against sea cliffs and the foundations of buildings. Examples of the resulting problems are examined in later chapters.

The most fearsome force of nature threatening the Northwest coast is the tsunami. In 1964 and 1968, tsunamis generated in Alaska caused some damage in Oregon and Washington, although the most extensive damage occurred in Crescent City, California. Extreme tsunamis generated by subduction earthquakes immediately off the coast have struck the Northwest coast in the past and pose the greatest threat to the coast in the future.

4 The Arrival of Man—Erosion Becomes a Problem

For countless ages the Northwest coast remained untouched by humans. The Pacific waves crashed against the high cliffs and sandy beaches, observed only by the abundant seals and sea lions that frequented the shoals of bays and ocean beaches. Their existence must have been idyllic until the day about 10,000 years ago that human beings first walked the shores of the Northwest coast. They were Asians from what is now Siberia, hunters who crossed over to Alaska on the land bridge that emerged when sea levels fell during the last ice age.

It is possible that humans first arrived on the Northwest coast in pursuit of game—not necessarily deer or elk, but larger game, the now-extinct mastodons. In the summer of 1977, a farmer near Sequim, on the south shore of the Strait of Juan de Fuca, turned up tusks and bones from a mastodon with his backhoe (Borden 1979). Firmly stuck into one of the ribs was the broken tip of a sharpened bone projectile, a clear indication of the presence of big-game hunters. Radiocarbon analyses of plant material from the site established that this ancient hunt took place about 12,000 years ago, soon after the glaciers had withdrawn from the strait.

The oldest material associated with Indians in what is now Oregon and Washington was found inside Fort Rock Cave in eastern Oregon and has been dated at about 13,000 years before the present. Still older material, dating back 15,000 years, has been found in Idaho. This supports archaeologists' hypothesis that the Indians moved down into the Northwest from Alaska by first moving eastward to skirt the still glaciated mountains. Accordingly, the general view is that the interior was settled first, the banks of the Columbia River next, and the coast last. The oldest dated Indian material found on the coast is about 8,000 years old. That age may be misleading, however, because earlier occupation sites directly on the coast would now be covered by the ocean, which rose at the end of the ice ages.

The Native Americans must have lived a reasonably carefree existence before Europeans arrived in the Northwest. Food was plentiful—berries and game on the land, salmon and shellfish in the sea. The native people used cedar for almost all of their material needs: clothing, shelter, utensils, containers, and their superb canoes. They lived in small villages of plank houses, generally on the shores of estuaries and streams rather than on open-ocean beaches. The coastal Indians were animists who believed that all things—rock and tree, animal and man—were imbued with spirit. And with this living world the Indians lived in close communion.

It is uncertain when Europeans first appeared on the Northwest coast, or even who they were. The first explorers probably arrived during the sixteenth century, and they probably were Spanish. Many historians credit a Spanish expedition that sailed in 1542 from Navidad, Mexico, under the command of the Portuguese explorer João Rodrigues Cabrilho (Juan Rodriguez Cabrillo in Spanish). Following Cabrilho's death, the expedition was directed by the pilot, Bartolomé Ferrelo. It is believed that they reached the vicinity of the Rogue River in the spring of 1543. They were prevented from landing, however, by storms so severe that the crew was assembled to make their last confessions.

Other explorers followed. One was the Greek mariner Valerianos, who sailed under the Spanish flag as Juan de Fuca. His course took him along the coast of New Spain and California, and he claimed to have sailed as far north as latitude 47 degrees, where he found the broad inlet to the sea that now bears his name. He saw people clad in animal skins and decided not to land.

The Spanish did not attempt to lay permanent claim to this new land. Their search for gold and wealth drew them instead toward the west, across the Pacific to the Philippines. For nearly 300 years there was extensive travel and trade between Spain's possessions in Mexico and the Philippines. The course followed by the great galleons was generally south of the latitudes of Washington and Oregon, but occasionally ships were blown off-course to the north. One of them, believed to be the *San Francisco Xavier,* wrecked at the base of Neahkahnie Mountain on the north Oregon coast, spilling its cargo of beeswax, remnants of which are still sometimes found on the nearby beaches. It is almost certain that Spanish treasure lies somewhere off the Northwest coast, perhaps buried deep within the beach sand, but none has ever been found.

The Spanish were not the only seamen to wash up as castaways on the shores of the Northwest coast, and they may not have been the first. Some castaways originated in the Orient, from China, Taiwan, and Japan. In their book *Ten Years in Oregon,* published in 1844, missionaries Daniel Lee and Joseph Frost wrote: "About 30 or 40 miles to the south of the Columbia are the remains of a vessel which was sunk in the sand near shore, probably from the coast of Asia" (quoted in Gibbs 1978). During a trip along that stretch of coast in 1848, Sol Smith and John Hobson reported seeing "several pieces of a junk" which they concluded was Chinese. Dozens of Asian junks have been found wrecked or adrift off the Northwest coast, and many were undoubtedly carried ashore, much as glass floats originating in the Orient still arrive during winter storms (Weber 1984; Plummer 1991).

Perhaps the most intriguing of the stories of Asian visitors is the Indian legend that about 500 years ago, just before Columbus discovered the New World, an old Chinaman survived the wreck of his junk on the shores of Oregon (Plummer 1991). The legend relates that he enlisted "bad" Indians, who pillaged and plundered one native village after another along the coast. The fierce Asian brought with him knowledge of how to make and wield weapons of metal, which were used with frightening results on their victims. Whether or not this legend is true, it foretold the later arrival of Europeans with their vastly superior capacity to wage war and subjugate the Indians.

Exploration of the Northwest coast began in earnest in the last quarter of the eighteenth century. Some of the explorers were still searching for the Northwest Passage, the mythical route that supposedly connected the Atlantic and Pacific Oceans and would provide a direct sea route from western Europe to Asia. However, the new objective was "the great river of the west," sometimes called the "Oregon" but now known as the Columbia (Nokes 1991). In August 1775, the distinguished Spanish explorer Bruno de Heceta located the river's mouth, but his crew were too weakened by scurvy to handle the sails and cross the bar. Two years later, the English mariner James Cook passed the river mouth unknowingly during a stormy night. He was followed in 1778 by his countryman John Meares, who concluded that there was no river at all, and so named its estuary Deception Bay and its northern promontory Cape Disappointment. While surveying the Northwest coast in 1792, Captain George Vancouver passed the river's mouth. Although he noted the presence of earth-colored water, drifting logs, and cross currents, he wrote in his log, "Not considering the opening worthy of more attention I continued our pursuit to the N.W." (quoted in Nokes 1991).

The Columbia River was finally found in 1792 by an American, Captain Robert Gray, on his second voyage to the Oregon coast. The American merchants of New England were anxious to trade with the Chinese, but they produced little the Chinese wanted. However, they had heard that Russian and British traders earned great profits by selling Northwest furs in the Orient. In October 1787, Captains John Kendrick and Robert Gray were sent out from Boston with a cargo of beads, cloth, and bits of iron and copper to trade for sea otter pelts. The pelts were sold in China, and the money used to buy tea and perhaps silk and spices. Having completed these transactions, Gray set sail for Boston, the first American to circumnavigate the globe.

On his second voyage to the Northwest coast, Captain Gray found himself off the mouth of a great river. Like Vancouver, he noted a large flow of muddy water fanning out from the shore. Even more so than now, the Columbia bar was one of the most treacherous on earth. At eight o'clock in the morning of May 11, 1792, after waiting several hours for the right combination of wind, tides, and currents, Gray gave the command and his ship, the *Columbia Rediviva,* crashed through the breakers and entered the Great River of the West. The Columbia River was finally discovered.

The next major thrust of exploration came from inland. On November 15, 1805, after an arduous 19-month journey, Meriwether Lewis and William Clark reached the Pacific Ocean at the mouth of the Columbia River. There they passed a miserable winter in a small log stockade, Fort Clatsop, built on a low hill above a bog and tidal creeks. It rained every day but six. There was much sickness—colds, dysentery, and rheumatism—and the men were plagued by fleas. Food was scarce. On Christmas they celebrated with "pore Elk, so much Spoiled that we eate it thro mear necessity, Some Spoiled Pounded fish and a fiew roots" (quoted in O'Donnell 1988). In spite of the adversity they faced, Lewis and Clark recorded a detailed account of the geography, fauna, and flora; they returned east with reports that this was a place suitable for settlement.

The first permanent community established by white men on the Northwest coast came soon after the departure of Lewis and Clark, and at the same locality—Astoria, on the south bank of the Columbia about 4 miles upriver from the mouth. John Jacob Astor was a German immigrant who had achieved the American dream of acquiring great wealth. One of his ideas was to establish a fur-trading center at the mouth of the Columbia. The furs brought in by Indians would be shipped out and sold in the Orient. In theory the idea was a good one, but in practice it was a disaster.

Two contingents were sent to establish the post, one by land, the other by sea. The *Tonquin* was captained by Jonathan Thorn, who turned out to be a psychopath. Due to his madness, eight men died while crossing the Columbia bar. It must have been something of a relief to everyone when Thorn was killed a few months later by Indians on Vancouver Island after he struck their chief. The overland contingent also was plagued by disaster. They became lost in the uplands of the Snake River and were reduced to eating their moccasins.

Matters did not improve much on their arrival at Astoria. They first had to clear a site for the trading post, and many of the trees in the surrounding forest were 50 feet in girth. After two months, barely an acre had been cleared, two men had been badly injured by falling trees, and one had blown his hand off; three of the company had been killed by Indians. In June 1812 the United States declared war on England. In the expectation that the British would arrive at any moment to seize the post, it was wisely sold to the British-owned North West Company. In December 1813 the Stars and Stripes came down and Astoria became Fort George. The British dominated the area for the next three decades, but a negotiated agreement eventually returned the mouth of the Columbia to the United States.

The arrival of white men inevitably meant the demise of the native inhabitants. Europeans brought disease, whiskey, guns, greed, and superior power. Disease was the first to strike, and it is certain that it arrived well before white men physically reached these shores in significant numbers. Some anthropologists and historians believe that the native population had already been reduced by half before Europeans arrived on the scene. The noted botanist David Douglas described the impacts of disease on the Indians, although at a much later date. During a visit to Clatsop County during October 1830, Douglas recorded the following entry in his journal: "A dreadfully fatal intermittent fever broke out in the lower parts of this river (the Columbia) about eleven weeks ago, which had depopulated the country. Villages which had afforded from one to two hundred effective warriors are totally gone; not a soul remains. The houses are empty and flocks of famished dogs are howling about, while dead bodies lie strewn in every direction on the sand of the river" (quoted in Gibbs 1978).

As white men arrived in increasing numbers, massacre as well as disease decimated the populations of Native Americans. An Indian agent recorded in his journal on February 5, 1854, that

> a most horrid massacre, or rather an out-and-out barbarous mass murder, was perpetrated upon a portion of the Nah-so-mah band residing at the mouth of the Coquille River on the morning of Jan. 28, by a party of 40 miners. The reason assigned by the miners by their own statements, seem trivial. However, on the afternoon preceding the murders, the miners requested the chief to come in for a talk. This he refused to do. Thereupon the whites at and near the ferry-house assembled and deliberated upon the necessity of an immediate atack upon the Indians. A courier was sent to the upper mines, some seven miles to the north, for assistance. Twenty men responded, arriving at the ferry house in the evening proceding the morning massacre.
>
> At dawn on the following day, led by one Abbott, the ferry party and the 20 miners, about 40 in all, formed three detachments, marched

> upon the Indian raches and consumated a most inhuman slaughter, which the attackers termed a fight. The Indians were aroused from sleep to meet their deaths with but a feeble show of resistance; shot down as they were attempting to escape from their houses. Fifteen men and one squaw were killed, two squaws badly wounded. On the part of the white men, not even the slightest wound was received. The houses of the Indians, with but one exception, were fired and entirely destroyed. Thus was committed a massacre too inhuman to be readily believed. (Quoted in Gibbs 1978)

In spite of such blatant atrocities, the Indians were slow to seek revenge, perhaps recognizing the hopelessness of their situation. Finally, on the night of February 22, 1856, the Indian tribes in the Rogue River area rose in desperation and went on the warpath. Practically every home inhabited by white settlers in Curry County was burned to the ground; many were killed. The Rogue River War of 1855–56 was a significant uprising of Northwest coastal Indians, but it came too late. It was soon put down by American troops, and the Indians who survived were herded onto reservations, whether or not they had participated in the uprising.

The disease and killing went on for more than a century, almost obliterating the native people from the Pacific shores. Their former existence is re-

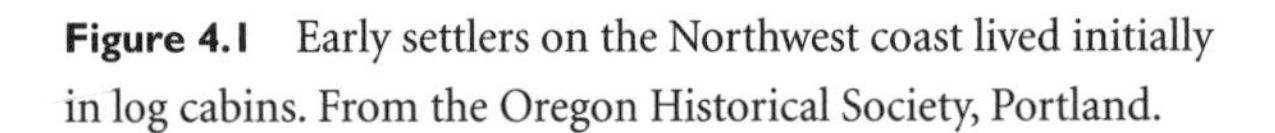

Figure 4.1 Early settlers on the Northwest coast lived initially in log cabins. From the Oregon Historical Society, Portland.

Figure 4.2 The initial settlement of Newport occurred along the shores of Yaquina Bay. Courtesy of Rose Troxel, Newport, Oregon.

corded in the geography of the Northwest coast: Yaquina, Siletz, and Willapa are derived from the names of Indian tribes. Their presence is also recorded in the great mounds of their shell middens—little evidence of a people who lived in relative peace in a plentiful land for thousands of years.

With the Indians out of the way, settlement by Europeans began in earnest. The first homes were crude cabins built of readily available logs (fig. 4.1). These were later replaced by more elegant dwellings. It is sometimes difficult to believe that all the development we see on the coast today—homes, shops, roads, bridges—has come about within the last 100–150 years. For example, attention was first drawn to Yaquina Bay in 1852 when the schooner *Juliet* was wrecked on the nearby ocean shore (Price 1975) and rescue attempts dramatized the bay's potential as a harbor. Settlers were slow to come to the area, however, and it was not until 1875 that the first plat was filed for the city of Newport. It covered a small area along the waterfront of Yaquina Bay, the present-day "old town" (fig. 4.2). During those formative years the town's day-to-day activities centered on the bay, and the same was true in towns and villages all along the coast.

"Summer people" began coming to the coast as early as the 1860s, arriving either by boat or by wagon on muddy roads that crossed the Coast Range. Like tourists today, they were drawn to the ocean beaches, where they set up tents and stayed for much of the summer. In Newport, the camping centered on Nye Beach (fig. 4.3). Completion of the railroad between Corvallis and Yaquina City in 1885, with ferry service to Newport, greatly increased the numbers of summer people.

Figure 4.3 Many of the "summer people" visiting Newport at the turn of the century stayed in tents along the shore at Nye Beach. Courtesy of Rose Troxel, Newport, Oregon.

Figure 4.4 As the number of visitors to Nye Beach grew, more permanent and comfortable accommodations were built, including a hotel and a sanatorium that provided hot seawater baths. Courtesy of Rose Troxel, Newport, Oregon.

Figure 4.5 The Burton expedition in July 1912 was the first automobile trip from Newport to Siletz Bay (see Stembridge 1975a). That arduous trip took 23 hours; today the 47-mile distance is easily covered in less than an hour. From the Burton Collection of the University of Oregon, Eugene.

The growing interest in beach recreation inevitably led to the construction of permanent buildings along the ocean shores. The tents at Nye Beach gave way to cottages and cabins, and stores were constructed to serve the growing community (fig. 4.4). In 1901 a bathhouse was built close to the shore at Nye Beach; the next year a sanatorium featuring hot seawater baths was constructed. And so went development all along the Northwest coast.

The railroads completed in the late 1800s facilitated east-west travel between the coast and the inland valley. North-south movement along the coast remained extremely limited, however, and coastal communities were isolated except for uncertain boat traffic between the bays or travel along the beaches at low tide. The problem of movement along the coast is illustrated by the Burton "expedition" of July 1912, which took 23 hours of arduous travel in a Flanders "20" car to cover the 47 miles between Newport and Siletz Bay (fig. 4.5; Stembridge 1975a). The eventual construction of Highway 101 along the coast was done piecemeal. The building of the Roosevelt Coast Military Highway began in 1919 with projects in Tillamook and Curry Counties (Boxberger and McFeron 1983). The section through Lincoln County was completed in 1922, but it was not possible to traverse the entire length of the coast on a quality road until 1932. Even then the numerous bays and estuaries had to be crossed on ferries. The magnificent bridges we

Figure 4.6 As indicated by the handwritten caption on this photo taken at Barview, at the entrance to Tillamook Bay, even early development along the coast was affected by erosion. From the Pioneer Museum, Tillamook, Oregon.

use today were completed during the 1930s. Only at that stage was the coast fully ripe for development.

The processes of wave erosion were readily apparent even in those early days of settlement (fig. 4.6), and it was inevitable that erosion would conflict with the growing pressure to develop land adjacent to the beaches. Houses, hotels, and roads inevitably were lost to the waves of winter storms. It might be a natural process, but coastal erosion did not fit in with human plans for development, and accordingly it became a "problem."

5 The Development and Destruction of Bayocean Spit

In 1906, T. B. Potter traveled from Kansas to Portland, Oregon, on vacation. Potter, a real estate promoter, had made a fortune as the head of the T. B. Potter Realty Company, which had offices in Kansas and San Francisco. In Portland he inquired about good places for fishing and hunting, and was advised to travel to Tillamook Bay on the coast (fig. 5.1). He found a large, beautiful bay teaming with fish and waterfowl. Beyond it lay a 4-mile-long peninsula covered with dunes and forest. Along its edge a wide, sandy beach met the surf of the Pacific Ocean. Potter's business instincts were aroused, and he vowed to develop the "Atlantic City of the Pacific Coast."

Potter's vision was the opening chapter of the first, and still the grandest, attempt at development on the Northwest coast. It was to meet with tragedy, first in the form of economic problems, and subsequently in the form of catastrophic erosion that ultimately led to the destruction of his resort.

The Development of Bayocean Park

The initial settlement of the Tillamook area had occurred less than a half century before Potter arrived there. Tillamook County deed records show land claims for portions of the peninsula, later to be called Bayocean Spit, dating back to 1867. One of the early landholders was A. B. Hallock, who constructed a wharf and warehouse on the bay side of the spit. In April 1891, Hallock established a post office in his home, primarily to serve the personnel at the Cape Meares lighthouse. He called his small holdings Barnegat, after a popular Atlantic coast resort.

Alight with his vision to establish an elegant resort on the spit, Potter returned to Portland and teamed up with H. L. Chapin to form the Potter-Chapin Realty Company. Potter purchased much of the spit, and in 1907 he submitted a plat map to Tillamook County. The map is an impressive display of some 3,000 lots, each 50 by 100 feet. Lots were to be priced from

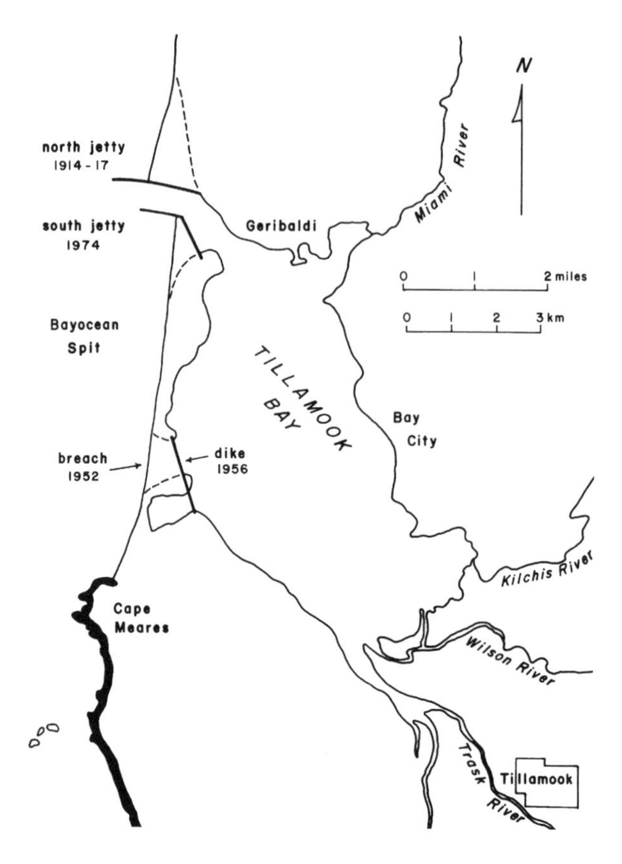

Figure 5.1 Bayocean Spit and Tillamook Bay today, with two jetties at the mouth of the inlet and a dike blocking the breach that developed in 1952 at the south end of the spit.

$150 to $1,800 depending on their view and proximity to the beach. A large block of land in the middle of the spit was reserved for a hotel and bathhouse.

Bayocean Park was the name chosen for the development because the ocean was on one side of it and the bay on the other. Optimism ran high. A June 29, 1907, account in the *Oregonian* described Potter's professional prowess:

> Mr. Potter has the reputation among his friends of never having touched an investment that has not forged to the front, an unqualified success. It is said that his greatest accomplishments are due to the fact that his observations are wonderfully discriminating; that he is quick to grasp opportunities for development and that his subsequent action is quick, well-directed and gets results. It is said of him that he has enabled more people to make more money in suburban realty than any other individual in the United States.

The article went on to describe the resort: "The most attractive vacation spot eclipsing all the resorts for tourist travel in Europe, excelling anything on the Atlantic Coast and outstripping the best on the Pacific shores is to be only a bit over a couple of hours ride from the very heart of this city" (Portland). The "couple of hours ride" referred to travel on the railway from Portland to Tillamook Bay that was expected to be completed by the next year. In the meantime, the trip took much longer.

The development plans for Bayocean Park called for a magnificent hotel set high on a hill overlooking the sea (fig. 5.2). Initially, however, only a three-story concrete-block structure that Potter called the "Hotel Annex" was built. Once the rest of the hotel was completed, this portion would be used to house servants. Even the Annex achieved a degree of elegance. It had

Figure 5.2 *Top,* the Bayocean Hotel as planned; *bottom,* the completed structure. From the Pioneer Museum, Tillamook, Oregon.

electric lights, steam heat, and automatic fire sprinklers. Its impressive kitchen and dining room specialized in native seafood.

The most interesting structure in Bayocean Park was the natatorium (fig. 5.3), which was built close to the beach and housed an indoor swimming pool (the developers recognized that the waters of the Pacific would be too cold for most home buyers). The sales literature described it as "the largest plunge in the world, 500 by 1,000 feet." Reality again came up short; the constructed pool was only 50 by 160 feet. However, the water was artificially heated, and an oscillating paddle at one end created surf. The sales literature described it as "the latest wonder, a wave-making device creating artificial surf—so real in its action as to fool old Neptune himself."

Figure 5.3 The natatorium contained a swimming pool with artificial surf generated by a paddle at one end. From the Pioneer Museum, Tillamook, Oregon.

An electric plant was installed to provide power for lights and equipment in the hotel, natatorium, and other public buildings. Drinking water was piped onto the spit from Coleman Creek on Cape Meares, more than 3 miles to the south.

Potter advertised lavishly and widely throughout the United States, and in spite of Bayocean's isolation, lots sold readily at first. Some 2,200 lots were sold at a gross income of more than $1 million. Houses began to appear on the spit. A few were elegant in their design, reflecting the wealth of the investors (fig. 5.4), but most were modest structures. There was also a "tent city" for those who could not afford more. A downtown of sorts came into being, with a grocery, bakery, cafés, and a rooming house, and, of course, an agate shop. A trapshooting gallery and bowling alley were also available. The old Barnegat Post Office was closed, replaced in February 1909 by the Bayocean Post Office. It was not long before Bayocean Park had year-round residents, requiring the construction of a small one-room schoolhouse.

The official grand opening of Bayocean Park was set for June 22, 1912. Potter should have had some satisfaction in the realization of his dream, but already the development was facing major problems. Most of these stemmed from its isolation. The railroad from Portland to Bay City, on the opposite side of the bay from Bayocean, planned for 1908 was not completed until the end of 1911, three years late. To overcome this problem, Potter bought a yacht, which he named the *Bayocean*, a sleek vessel with clean lines. The yacht was to bring passengers from Portland, down the Columbia River, and then across 60 miles of ocean to Tillamook Bay. This did not prove to be an effective scheme. The trip sometimes took two or three days, and the passengers arrived at Bayocean Park weak from seasickness.

Figure 5.4 A group of small but elegant houses atop a hill on Bayocean Spit early this century. From the Pioneer Museum, Tillamook, Oregon.

This early isolation prevented lots from selling at the hoped-for rate. The resultant lack of capital made it impossible to build some of the promised improvements and created dissension among the investors. With the completion of the railroad Bayocean Park should have taken on renewed life. However, this was the age of the automobile, and tourists preferred to drive to resorts such as Seaside. Promised improvements at Bayocean still were not forthcoming. Complaints increased, culminating in charges that the enterprise was bankrupt. A petition was initiated in 1914 for a receiver. Potter fought the charges, but the plaintiffs' evidence must have been convincing because the court appointed a receiver. Years of litigation and claims of fraudulent practices followed.

In 1927, a group of prominent Tillamook residents organized as the Tillamook-Bayocean Company to assume the entire assets of the T. B. Potter Company's holdings on Bayocean Spit. Much of the direction for this new group came from F. D. Mitchell, the first purchaser of property on the spit. A road was finally completed along the south side of the bay and out onto the spit, and the future of the resort again brightened, but only briefly. The Great Depression of 1929 destroyed all hopes for a revival of the development. One by one the cottages were abandoned; the hotel closed.

The last true community on the spit consisted of a commune of drifters and other products of the Depression who lived in the hotel and fished and caught crabs to sell to the surrounding communities. Their efforts came to the attention of the government, and the Federal Emergency Administration granted the colony $3,900 to purchase fishing equipment and building materials. They were also given possession of a government-owned 65-foot cutter to be used in tuna fishing. The colony appears to have flourished for a short time, but then it disappeared.

One last glimmer of hope for Bayocean Park came in 1936 with the opening of the Wilson River Highway, which provided an easy route for automobile travel between Portland and Tillamook. Real estate men began approaching the Bayocean investors with discouraging reports about the resort followed by offers of small sums of money. Unfortunately, the discouraging reports were true. Not only had the resort sadly deteriorated, it now also faced increasing erosion, which was undermining and destroying the community.

Construction of the North Jetty

Nearly concurrent with the completion of the railroad and the official opening of Bayocean Park came approval for the construction of a jetty at the entrance to Tillamook Bay. The jetty's construction and the resulting erosion provided the final blow to the already tottering development.

There was some interest in "stabilizing" the inlet to the bay as early as the 1880s, but it intensified in the first decade of this century when Tillamook experienced an economic upswing and prospects for the community looked bright. Local industrialists, led by lumber interests, lobbied state and federal politicians to "improve" the Tillamook channel to create a deepwater port. Initially the U.S. Army Corps of Engineers was opposed to the jetty on the basis that there was no need for improvement. A Corps report from the 1880s stated:

> It is believed that the development of Tillamook Bay commerce is in no way hindered by the present condition of the bar and entrance, in that any vessel that can navigate the bay will find no difficulty in crossing the bar. Tillamook bar has always been considered by mariners as one of the safest along the coast on account of its generally uniform depth and position, and no wreck is known to have ever occurred in crossing it.

By 1903 the Corps had come around to backing jetty construction and introduced the first design:

> After careful study of the map and local conditions, the following project for obtaining 15 feet at mean lower low water is submitted. It involves the construction of a north high tide jetty of rubblestone from the permanent North Head near Green Hill, running seaward in a gentle curve concave to the ebb. This jetty would act partially to prevent the greater part of the sand movement on the north side, but chiefly as a retaining wall to gently control the ebb current and to keep it from spreading out to the north, and to confine it between this jetty and the shoal sand spit. . . . At the same time it would seem desirable to build a shorter high-tide south jetty extending out from Kincheloe Point 4,400 feet from high water mark, to check the cyclic sand movement into the harbor from the south.

Congress rejected the proposal, probably because of Tillamook's proximity to the port of Astoria on the Columbia River. The upswing of Tillamook's economy in 1907–8 prompted another consideration, but the proposal was again turned down. Congress indicated, however, that the proposal would be reconsidered after the long-hoped-for railroad from Portland was completed and the anticipated increase in commerce and shipping traffic justified channel improvements.

Approval for the jetty's construction finally came in 1911, but a new problem arose: the federal government would pay only half the cost; the cities of Tillamook and Bay City had to pay the other half, roughly $1 million. The small communities were unable to raise that much money, so the commissioners of the port of Tillamook proposed that only the north jetty be constructed, thereby decreasing the cost (Weber and Weber 1989). The Corps pointed out that the action of a single jetty cannot be predicted and noted that experience elsewhere on the Northwest coast demonstrated that both jetties were needed. Economics, however, dictated that only the north jetty be constructed, and economics won the day.

Work commenced in June 1914. Rock was quarried near the Miami River and hauled 7 miles by train to the north jetty site. The jetty was completed in October 1917, having utilized a total of 429,000 tons of rock at a construction cost of $766,000 (Terich 1973). The new jetty was 5,400 feet long and projected seaward with a slight curve from the base of the hills on the north side of the channel (fig. 5.5).

The Erosion of Bayocean Spit

Changes in the inlet channel and adjacent shorelines were soon evident, particularly immediately north of the jetty, where sand accumulated so fast that the shoreline moved out almost as rapidly as the jetty was built (figs. 5.5

Figure 5.5 The inlet to Tillamook Bay: *left,* in October 1902, before construction of the north jetty in 1914–17; and *right,* in September 1941. The natural channel was broad and deep, offering reasonably safe passage to boats, but construction of the jetty produced a deep, narrow channel tight against the jetty and a large shoal offshore from the tip of Bayocean Spit. From Komar and Terich 1976.

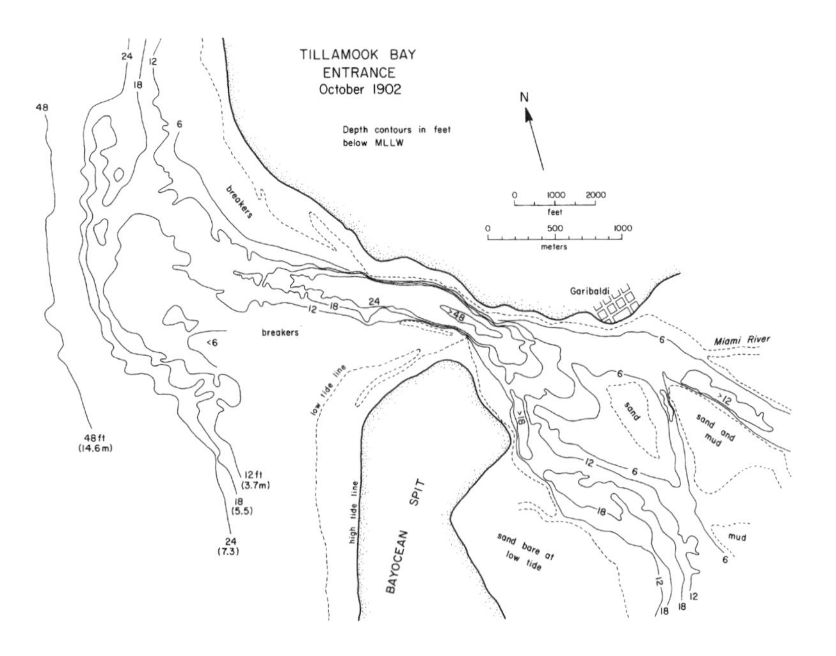

and 5.6). However, the impoundment of sand slowed and finally stopped once the shoreline was a smooth continuation of the shore farther to the north and away from the immediate effects of the jetty. By 1920, sand had ceased to accumulate north of the jetty, and the position of the shoreline was much the same as it is now.

South of the jetty, on Bayocean Spit, the impact was felt more gradually. It is difficult to establish when noticeable erosion began following completion of the north jetty in 1917. The initial changes may have been primarily offshore. Later comparisons of bathymetric charts revealed that the water progressively deepened during the 1920s (Dicken 1961; Terich 1973; Terich and Komar 1974). The direct impact on Bayocean Park itself began to become evident in the 1930s. Photographs of the natatorium that appeared in the *Oregonian* in 1932 and 1936 show the progress of the erosion (fig. 5.7). The 1932 photograph shows the sidewalk fronting the natatorium undermined and nearly gone; by 1936 the roof had caved in.

A resident on the spit used stakes to mark the progress of the erosion over a 7-year period and found that the rate of retreat of the foredune averaged 1 foot per year from 1926 to 1931 (Dicken 1961). The rate shot up to 6 feet per year during 1932 and 1933. This increase coincided with repairs on the north jetty and its lengthening by 300 feet. The perception of the residents of Bayocean Park was that the jetty lengthening caused the increased erosion.

The high rate of erosion along the spit continued through the 1930s and 1940s. Erosion losses were particularly severe during January 3, 4, and 5,

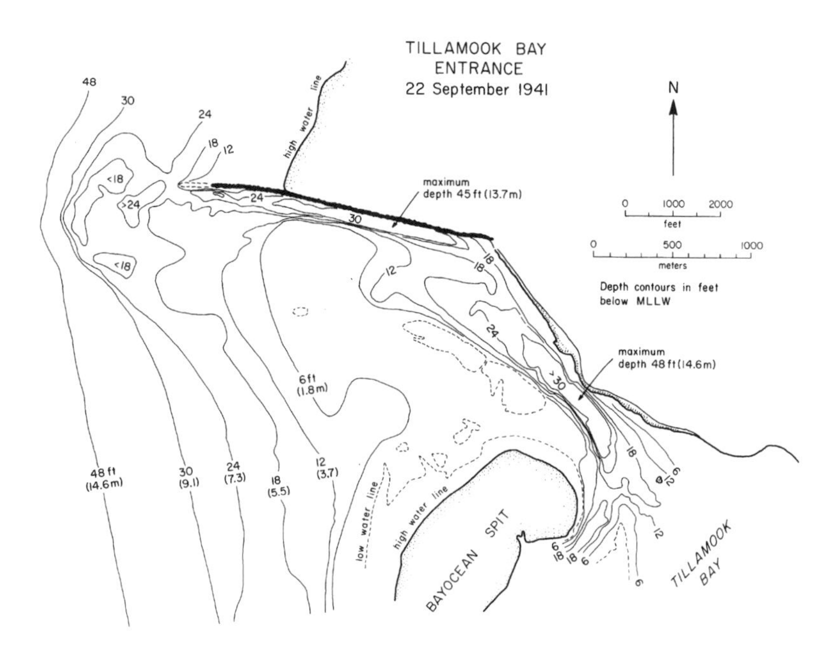

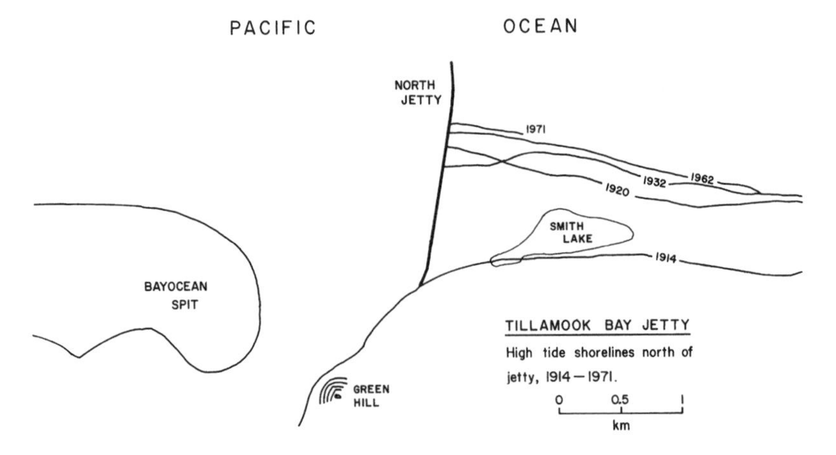

Figure 5.6 High-tide shorelines north of the Tillamook Bay jetty, 1914–71. From Terich 1973.

1939, when a storm combined with high tides to attack the spit. Small gaps were awash along the narrow south end of the spit, and sand and gravel were carried into the bay. The access road and water supply were cut off for a time. The last vestiges of the natatorium were destroyed, leaving only its foundations visible on the beach at low tide (fig. 5.7). The maximum recession of the dunes during January and February 1939 was 25 feet. Four houses had been undermined and destroyed, nine were immediately threatened, and six more had to be moved back to escape the erosion.

Storm waves returned in January 1940, cutting back the foredunes still further and widening the gaps in the narrow part of the spit. A group of citizens organized the Bayocean Erosion Committee to lobby government officials to come to their aid. The committee petitioned state and federal representatives but received little support. The war years postponed discussion of any possible aid for Bayocean Park. In 1946 the Corps of Engineers appraised possible measures to halt the erosion but concluded that any reasonable schemes would either have little effect on reducing the erosion or could not be economically justified.

The prospects for Bayocean Park should have been dismal. Most of the homes had been lost, and the remains of the once-grand hotel had reached the edge of the eroding bluff (fig. 5.8). Yet a 1947 newspaper article described a revival of the resort:

> Judging from the way people are buying property at Bayocean it will not be long before most of the desirable and usable lots which the County has long carried on the books for taxes will be back on the tax rolls as an asset to the County as well as the taxpayers. The old belief that Bayocean has been carried out to sea is being exploded and the more hardy souls who take time to look around and see what an enor-

Figure 5.7 The natatorium on Bayocean Spit in 1910 (*top*), c. 1932 (*middle*), and 1940 (*bottom*). By 1940 only its foundation remained. From the Pioneer Museum, Tillamook, Oregon.

Figure 5.8 The once elegant Bayocean Hotel succumbs to erosion during the 1940s. From the University of California, Berkeley.

> mous pile of sand is still to be washed away to eradicate Bayocean have come to the conclusion that it will be a long time if there is any further erosion at the rate that has been eroded in the past. (*Headlight Herald*, September 7, 1947)

The optimism must have been short-lived. Major erosion returned in November 1948, cutting new gaps through the narrow portion of the spit. Once again the access road was undermined and utility lines severed. The residents on the spit were isolated from the mainland, and food and water had to be ferried to them. The 1948 erosion effectively spelled the end of Bayocean Park, reflected in the title of an article that appeared in the January 2, 1949, issue of the *Oregonian*: "A Queen Dies: Victim of Time and Tide."

The final breaching of Bayocean Spit came on November 13, 1952, when storm waves aided by high tides penetrated the already weakened gaps and quickly cut away a 4,000-foot-wide section (fig. 5.9). Even at low tide the gap remained flooded with water. Most of the residents of Bayocean Park decided to leave and were evacuated by boat. Six chose to remain, and a system of bonfire signals was established between the residents on the spit and the Coast Guard at Garibaldi, to be used in case of an emergency.

Over the next four years the waves and tidal currents widened the breach. At the same time, the original bay inlet to the north shoaled and ap-

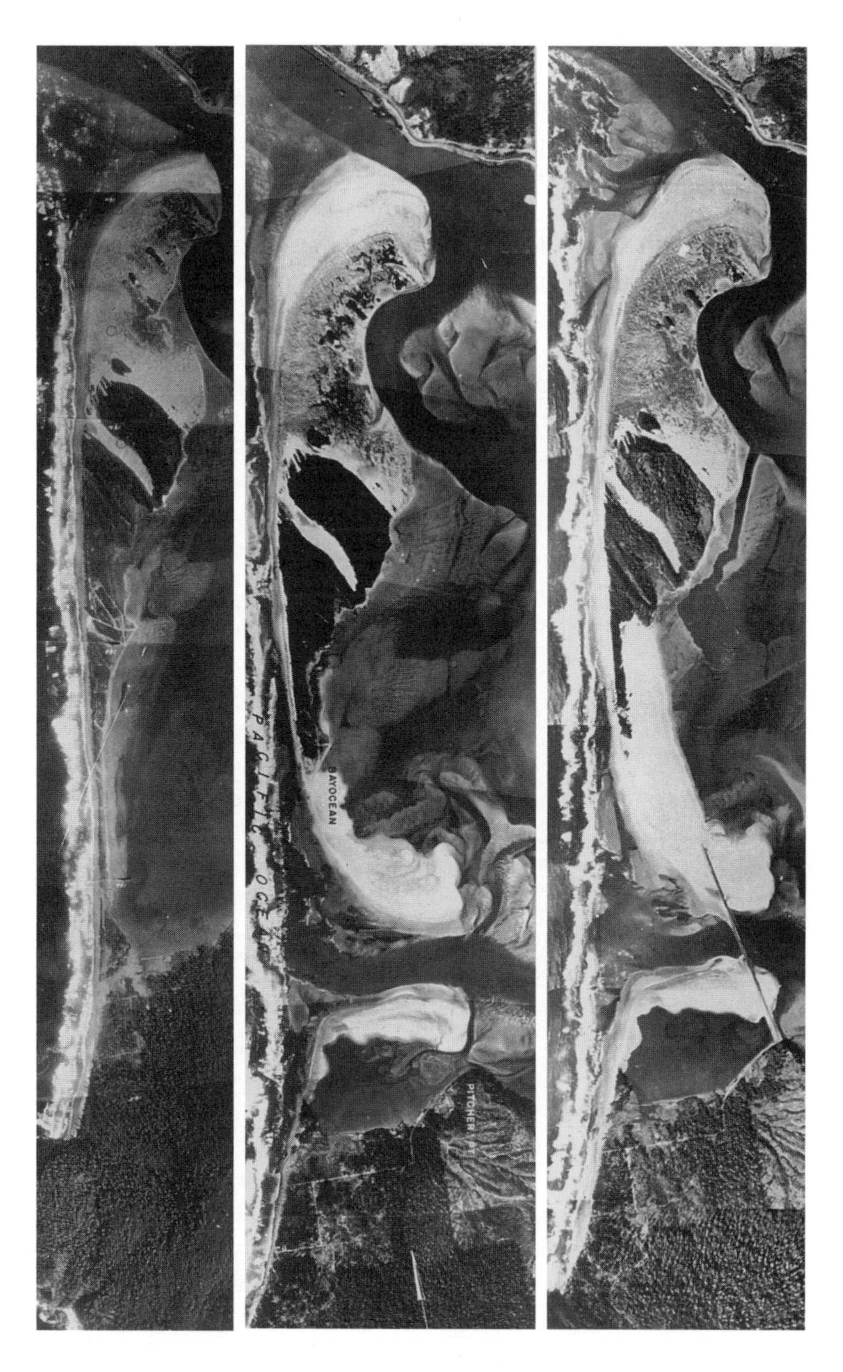

Figure 5.9 A sequence of aerial photographs of Bayocean Spit taken by the U.S. Army Corps of Engineers on September 26, 1945, April 4, 1956, and October 1, 1956, shows the progressive thinning of the spit due to erosion, its breaching in 1952, and the construction of a dike to close the widened breach in 1956. Courtesy of U.S. Army Corps of Engineers, Portland District.

peared to be on the way to closing altogether. The 1952 breach was developing into the natural opening for Tillamook Bay. This created additional problems. Waves rolled through the breach and caused erosion of farmlands along the south edge of the bay.

Plans were developed as early as 1953 to close the breach with a rubble-mound dike (Brown et al. 1958). The cost was placed at $1.8 million. The dike was approved in 1954, but construction was delayed until the summer of 1956. It was recognized that the dike had to be completed in one summer season because a partially completed dike would be severely damaged by winter storms. The alignment of the rock dike was set back from the ocean beach in the expectation that the pocket created in the shoreline would fill with beach sand, forming a more or less uniform shoreline (fig. 5.9). This proved to be the case. The new beach consisted of a ridge that trapped a small body of water (Biggs Cove) between it and the dike. Immediately following closure of the breach, fishermen reported a marked increase in tidal action within the navigation channel next to the north jetty and the rapid erosion of sandbars that had accumulated there.

Patterns of Erosion Due to Jetties

There are many examples from around the world of beach erosion resulting from the construction of jetties (Komar 1997). It is well established that significant erosion usually occurs when jetties block a net littoral drift of sand. In almost all cases the jetty acts like a dam blocking the longshore drift of sand, and the shoreline accretes on the updrift side while erosion occurs on the downdrift side (see chapter 3). For example, jetties built in 1935 to stabilize the inlet south of Ocean City, Maryland, trapped a strong southward littoral drift of sand. The shoreline advanced considerably on the north side of the jetties, opposite Ocean City, and eroded to the south along Assateague Island. The erosion was so severe that for a time it actually flanked the landward end of the south jetty.

The initial analyses of the erosion at Bayocean Park indicated that it fit the usual pattern created by jetties blocking a net littoral drift (U.S. Army Corps of Engineers 1970). And indeed, at first glance, the sand accumulation and shoreline advance north of the jetty (figs. 5.5 and 5.6), together with the simultaneous erosion along most of the length of the spit to the south, do suggest a net sand transport southward. On the basis of the rate at which sand was accumulating north of the jetty, the Corps estimated that the southward littoral drift was nearly a million cubic yards per year, just about the highest anywhere in the world.

But there is a problem with this interpretation. Just to the south of Bayocean Spit is Cape Meares, a large headland that should also have blocked southward movement of sand along the beach—if it existed. In fact, there is no built-out sand beach north of Cape Meares, only gravel and cobbles. We now know that this stretch of ocean shore, like others on the Oregon coast, is effectively a pocket beach isolated between headlands. Although there are seasonal reversals in the directions of sand movement along the shore, the long-term average is zero; that is, there is no net littoral drift. At one time, engineers believed that erosion problems do not develop in response to jetty construction on coasts where there is no net longshore movement of beach sand. The jetty construction at Tillamook Bay and subsequent erosion of Bayocean Spit demonstrated otherwise.

Like the north jetty at the inlet to Tillamook Bay, most of the other jetty systems on the Oregon coast were built early this century and, to varying degrees, also caused beach erosion. However, there was minimal development of the coast in those early years; only at Bayocean Park were houses and other structures in the path of the erosion. The shoreline changes associated with those other jetties are of interest because they establish the pattern of shoreline changes that occur when jetties are constructed on a coast where there is zero net littoral drift.

The events that followed the construction of the jetties at the mouth of the Siuslaw River (1891–1915) are particularly enlightening (Komar et al. 1976a). Especially large changes in the shoreline occurred north of the north jetty, as indicated by the seaward shift of the high-tide shorelines shown in

Figure 5.10 Shoreline changes at the mouth of the Siuslaw River near Florence on the mid-Oregon coast caused by jetty construction early in the century. The 1889 shoreline predates construction of the jetty. Accretion occurred immediately north and south of the jetties; the new sand came from beach erosion farther along the coast. From Komar, Lizarraga, and Terich 1976c.

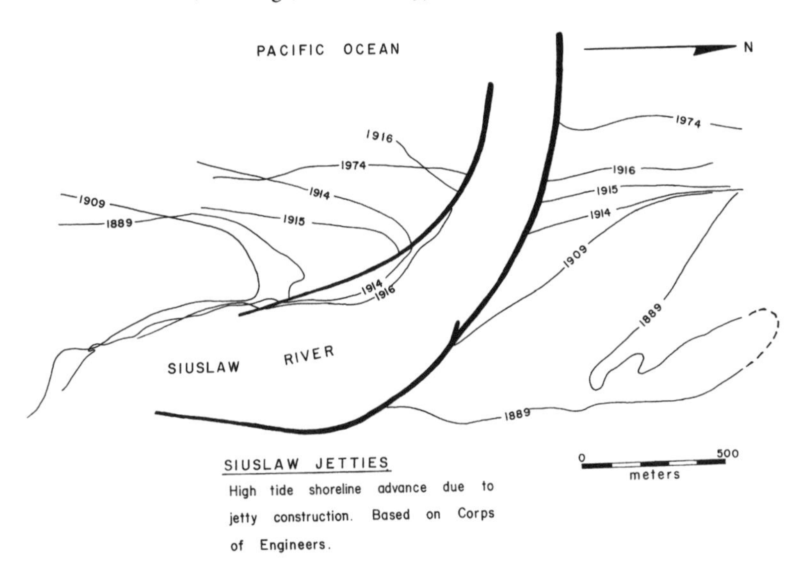

figure 5.10. The 1889 shoreline is given as the prejetty position, and it is apparent that the construction of the north jetty produced a large embayment. The compilation of the shorelines shows that this embayment progressively filled with sand, and that the beach migrated out as the jetty was constructed. The embayment filled with sand out to the point where the shoreline was a smooth continuation of the shore farther to the north, and then stopped. Aerial photographs of the 1974 shoreline indicate that it is nearly the same as the shoreline in 1939 aerial photos, the oldest ones available for the area. Thus, a new postjetty equilibrium had been achieved by 1939, and there were no further adjustments after that. There was some accretion south of the jetties, but the volume was smaller than that to the north because the embayment created by the south jetty was smaller. The shoreline there has also been stable since at least the 1930s.

Before the jetties were built, sand movement at the mouth of the Siuslaw inlet must have been in equilibrium; waves carrying sand toward the mouth must have equaled the tidal currents and river flow moving sand away. The equilibrium state was a shoreline curving inward toward the mouth. A disequilibrium was created when the jetties were constructed. The waves broke at an angle to the curved shoreline and carried sand into the embayment, where it accumulated. This continued until the shoreline again became parallel to the wave crests and the net sand transport was again zero. At that stage, probably achieved sometime in the 1920s in the case of the Siuslaw jetties, no additional readjustments of the shoreline were required since the beach was once again in equilibrium with the wave climate.

The patterns of shoreline changes associated with the construction of the north jetty at the Tillamook Bay inlet were similar (fig. 5.11). Beach sand accumulated immediately north of the jetty but eventually halted when a new equilibrium was achieved. Had there been a net littoral transport to the south, the shoreline advancement would not have stopped as observed, at least not until it reached the end of the jetty, at which time the southward sand drift would have spilled around the end of the jetty. Although erosion did occur along most of Bayocean Spit, closer inspection showed that some sand deposition took place at the north end of the spit next to the inlet. In the absence of a south jetty, the accumulating sand formed shoals within the mouth of the inlet rather than developing as a simple shoreline advance (fig. 5.12). Therefore, rather than aiding navigation through the inlet into the bay, construction of the north jetty produced sand shoals and a larger bar, which greatly increased the hazards.

The fact that initially only the north jetty was built at Tillamook undoubtedly increased the erosion of Bayocean Park. In the absence of a south jetty, sand from the beach opposite the resort continued to be fed into the shoals at the inlet mouth, with some of it carried into the bay by tidal currents. With this arrangement, no new equilibrium could be established with

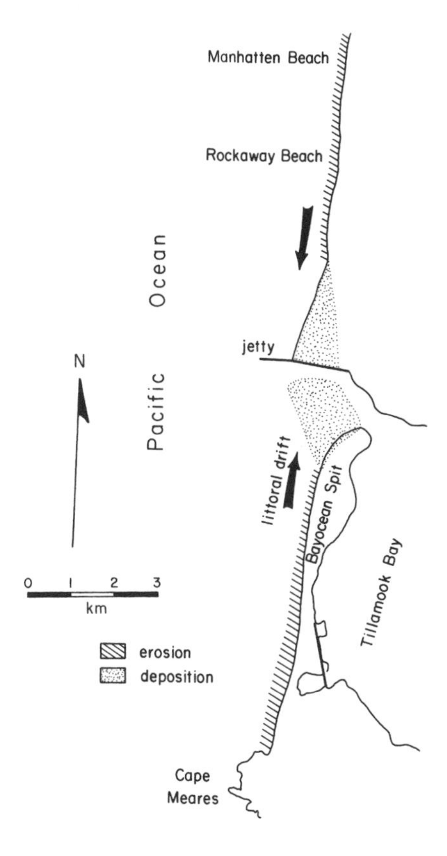

Figure 5.11 Patterns of erosion and sand deposition in response to the construction of the north jetty at the Tillamook Bay inlet. Sand derived from the erosion of Bayocean Spit accumulated to form an extensive shoal at the mouth of the inlet. From Terich and Komar 1974.

Figure 5.12 The entrance to Tillamook Bay on May 22, 1947, photographed at low tide and in calm seas. Shoals have built out into the entrance from the north end of Bayocean, seen to the right. These shoals were covered by breaking waves during high tides, making the entrance extremely hazardous to ships. Courtesy of U.S. Army Corps of Engineers, Portland District.

Figure 5.13 Bayocean Spit on September 5, 1974, after construction of the south jetty. Sand accumulated immediately south of the jetty so that a pocket beach formed between the jetty and Cape Meares.

a zero-drift shoreline orientation to the south along the spit. Fortunately, a south jetty was eventually authorized, and it was completed in November 1974 (fig. 5.13). Sand quickly accumulated to its south, supplied by additional erosion of Bayocean Spit. Within a few years the sand accumulation next to the jetty halted, as did the erosion along the rest of the spit that supplied sand to fill the embayment created by the jetty. The shoreline along the length of Bayocean Spit has, after nearly a half century, once again achieved an equilibrium orientation in which the net longshore sand movement is zero. The symmetrical pattern of erosion and deposition on opposite sides of the Tillamook Bay jetties is now more like the patterns seen at the other inlets on the Oregon coast where two jetties were constructed at the start.

Summary

When T. B. Potter developed a grand resort on the dune-covered sand spit between Tillamook Bay and the Pacific Ocean, he vowed to construct the "Atlantic City of the Pacific Coast." Initially Bayocean Park appeared to have achieved Potter's vision. Lots sold well, a hotel and natatorium were built, and a number of homes were built among the dunes backing the beach.

Figure 5.14 Building foundations and rusted pipes littering the eroded beach are all that remain of Bayocean today.

But in the end, Bayocean Park became the first major development on the Northwest coast to fail. Its final demise was brought about by erosion, which cut away at the ocean beach and resort properties. All that remains today of the "Atlantic City of the Pacific Coast" are building foundations and rusted pipes littering the eroded beachface (fig. 5.14). The cause of the erosion was initially misunderstood to be blockage of a littoral drift of sand along the beach, a problem experienced on many other coasts. But the erosion on Bayocean Spit was unusual, and hence of special interest to coastal scientists and engineers. We once believed that jetties could be safely constructed on coasts that have no net longshore sand movements. The erosion of Bayocean Spit demonstrated otherwise.

6 Natural Processes of Erosion

The destruction of Bayocean Spit can be attributed to the construction of a jetty at the entrance to Tillamook Bay. In this case, humans can be "credited" with having caused the problem. Much of the beach erosion along the Northwest coast is not the result of human activities, however, at least initially.

Here we will examine the natural processes of beach erosion that periodically attack the shore. These processes are more complex than might initially be assumed. Severe beach erosion involves more than a major storm producing big waves. It requires a combination of unusually intense waves, nearshore currents that cut back portions of the beach, and high water levels produced by tides, wave setup, and storm surge. Generally, all or most of these factors must act in concert to cause erosion sufficiently severe to result in major property losses.

Natural processes of erosion occur along the full length of the Northwest coast, but they are most dramatic—and do the most damage to property—on sand spits, and they have been studied mainly in that context. We will first examine the erosion that has taken place on Siletz Spit south of Lincoln City on the Oregon coast. The problems experienced there exemplify the natural processes of erosion. We will then examine other cases of erosion; some reaffirm the findings at Siletz Spit, while others illustrate additional processes and problems.

The Development and Erosion of Siletz Spit

Development and Erosion Problems

Before the mid-1960s, Siletz Spit appeared much as it had for hundreds of years. The spit separated the ocean from Siletz Bay and had a narrow connecting inlet at its north end (fig. 6.1). The spit itself was covered with low,

Figure 6.1 Siletz Spit, a narrow stretch of land covered by low dunes, separates the Pacific Ocean from Siletz Bay except for the inlet at the north end of the spit. In 1971, when this photo was taken, only a few homes had been built on the spit.

Figure 6.2 A general view of the homes on Siletz Spit in 1971. The houses rest on a small bluff in the foredunes that represents the scarp of the eroded foredunes formed in the early 1960s, just before the spit was developed.

hummocky dunes and sparsely vegetated with dune grasses and wild strawberries. The coarse-sand beach sloped steeply toward the sea. Sea lions were its most common visitors.

The development of Siletz Spit began in the early 1960s. A road was constructed along its length and artificial lagoons were carved into the back side of the spit. A scatter of houses appeared, some atop the low foredunes immediately backing the beach (fig. 6.2). Problems with beach erosion began soon afterward. During the winter of 1970–71, a series of storms cut into the dunes and threatened the homes closest to the beach. The erosion was not particularly severe, and the dumping of loose rock (rip-rap) in the area of concentrated wave attack halted the dune retreat. However, this event was a warning of potential problems on the spit. Recognizing this potential, I submitted a proposal to Sea Grant to undertake a study of the hazards. Almost immediately, major erosion began and my investigation was under way in earnest.

A major storm struck the north Pacific just before Christmas 1972, generating huge breakers all along the coast (see chapter 3 and fig. 3.7). The waves quickly cut into the foredunes of Siletz Spit, threatening a group of houses midway along the spit. One of the houses was still under construction, and no attempt was made to protect it from the advancing surf. Within a week the eroding dune bluff reached the house, which collapsed onto the beach (fig. 6.3). A frantic attempt was then made to save the neighboring houses by dumping rip-rap along the eroding dunes. Initially only the seaward sides of the homes were protected, and the empty lots next to them were left exposed to the waves (fig. 6.4). Dune erosion continued on the unprotected lots and flanked the rip-rap fronting the houses. The owners had to add more rip-rap along the sides of their homes; several houses ended up on rock promontories extending into the surf (fig. 6.5). In less than three weeks, the dunes along the shore of Siletz Spit retreated nearly 100 feet.

The area of maximum erosion was directly seaward from an artificial lagoon cut into the back side of the spit (fig. 6.4). This raised fears that if the erosion reached the lagoon, the entire spit would be breached, much as had occurred at Bayocean 20 years earlier. The erosion received widespread newspaper coverage, and Governor Tom McCall brought together a group of scientists and engineers to advise him on what might be done to save the spit. However, the political winds shifted and blocked the state's rescue of the homeowners. Public access to the area is limited by a gate across the road leading into the development, and the public was not enthusiastic about spending state funds to help solve private erosion problems. The committee of scientists and engineers documented the nature of the erosion problem and determined how it could be halted. But they also pointed out that large quantities of mud and logs had accumulated in Siletz Bay immediately behind the eroding area, and that, loss of homes aside, a breach

Figure 6.3 A house under construction on Siletz Spit was lost during the 1972–73 erosion: *top,* December 28, 1972; *bottom,* January 19, 1973.

might help to flush out the bay and improve its circulation. Accordingly, Governor McCall backed away from the problem, wisely leaving the developer and homeowners to deal with it.

Another difficulty arose in that the lots on Siletz Spit are not purchased outright but are leased from the developer on a 99-year basis. This led to a disagreement over who should pay for the placement of the rip-rap: the developer who technically still owned the lots, or the individual leaseholders who owned the homes or were making payments on still empty lots. With

Figure 6.4 Houses threatened by erosion during the winter of 1972–73 were protected by rip-rap, but the undeveloped lots were left unprotected and continued to erode. Photo A514-65 from the Oregon Highway Department, January 4, 1973.

their houses in the path of the erosion, the homeowners could little afford a long argument and so quickly covered the costs of rip-rap placement. But the empty lots continued to erode. Individuals were not inclined to invest the thousands of dollars required to protect vacant lots, and in many cases discontinued their lease payments, with responsibility reverting to the developer.

Erosion on the empty lots persisted until it came within a few feet of the road leading out onto the spit and the utility lines buried along the side of the road. Only then, after the empty lots had effectively disappeared, did the developer step in and order rip-rap protection. The erosion was ultimately halted just short of the road.

Although the erosion was rapid and dramatic, only one home was lost during the winter of 1972–73, the house that had been under construction. If any lessons were learned, however, they were short-lived; development on the spit continued. The first step was to restore the lots that had been lost. This was accomplished by scooping sand from the beach and dumping it back onto the lots, building them up to their former levels. Soon thereafter another storm struck, causing some minor erosion of the restored lots but primarily littering them with drift logs hurled up by the surging waves. Log movements during storms are a severe hazard to homes built along the beach edge. Eventually homes were constructed on these restored lots.

Over the ensuing years, erosion of Siletz Spit has varied in its intensity from one winter to the next. There was major erosion during February 1976 and again in February 1978 (fig. 6.6). In each instance, more rip-rap was added. Eventually, rock armor came to cover nearly the entire spit. The rip-rap has served its purpose in protecting the homes, but in many areas the natural sand has been replaced by a mound of rocks.

Processes of Erosion

The erosion of Siletz Spit resulted from a combination of ocean processes. Most obvious and critical were the major storms that generated exceptionally high waves. The storm system that produced the December 1972 erosion covered most of the North Pacific. Although the storm winds themselves were not extreme, their fetch extended across thousands of miles of ocean surface, directed straight at the Northwest coast. The tremendous fetch and long duration of the storm permitted the transfer of an enormous amount of energy to the developing waves. When they finally reached the coast, the waves broke with awesome force against the shore, achieving significant wave heights on the order of 23 feet (see fig. 3.6). In contrast, the storm of February 1976, which also produced erosion, was characterized by extreme gale force winds (60–70 miles per hour) but a much smaller fetch. Important to wave growth in this case was the rapid movement of the storm across the Pacific. The storm kept pace with the moving waves, and the storm winds continuously pumped energy into the growing waves. When the waves reached the Oregon coast, they were nearly as high as those produced by the 1972 storm.

During each erosion episode, the maximum period of beach erosion corresponded to the largest storm wave heights, demonstrating that this

Figure 6.5 As the adjacent unprotected lots continued to erode, this house was left on a promontory of rip-rap extending into the surf. The foundation on the beach is from the house pictured in figure 6.3.

Figure 6.6 Rip-rap was placed on Siletz Spit in March 1976 to protect homes from erosion.

must be the primary cause. However, there were significant variations in the severity of property losses along the spit. The erosion was centered on a dozen or so lots and homes. Farther along the spit there was virtually no erosion in spite of the intense storm waves. The variability in the amount of the erosion was controlled by the presence of rip currents.

As described in chapter 3, rip currents can hollow embayments into the beach that bring storm waves nearer to back-beach properties. This process is particularly important on Siletz Spit because of the coarseness of the beach sand. Rip embayments are particularly deep and narrow on steep coarse-sand beaches. The rips and their embayments commonly precede the erosive storm waves rather than being produced by them. In this sense they set the stage for erosion, acting to direct the wave attack when a storm does occur.

This situation was apparent during air flights over Siletz Spit at the time of its erosion during the winter of 1972–73 (fig. 6.7). An extensive rip embayment was located midway along the spit, and it had cut entirely through the beach to reach the developed lots. The houses bearing the brunt of the erosion at that time, including the one that was lost (figs. 6.3 and 6.5), were in the direct lee of this embayment. When the storm of December 1972 struck, only this immediate area of the rip embayment suffered extensive property erosion. Rip embayments were visible elsewhere along the spit (fig. 6.7), but they had not developed to the extent that they reached back to the foredunes and houses.

The positions of the rip currents change from one winter to the next, concurrently shifting the areas subjected to maximum erosion, and it may be several years before erosion returns to specific lots. We have been unable to explain or predict the positions of rip currents and the embayments they create. However, their observed locations during the late fall and early winter can predict likely zones of property erosion, should a major storm occur.

Still another factor in sand spit erosion is the water level along the coast at the time of the storm. Recall from chapter 3 that the water level is determined by the combined effects of tides, the mean sea level, and the presence or absence of storm surge and wave setup. The role of tides is fairly obvious, it being apparent that the higher waters associated with spring tides will raise the shoreline closer to or higher up on coastal properties, thereby enhancing erosion. Neap tides prevailed during the December 1972 storm on Siletz Spit, whereas tides were nearly at spring-tide levels during the February 1976 storm. It was during the higher water levels of 1976 that drift logs

Figure 6.7 Rip current embayments cut through the beach, reaching the foredunes and development on Siletz Spit during the winter of 1972–73. The largest embayment (*top*) is the center of the property losses photographed in figures 6.3 and 6.5.

were thrown atop the restored lots on the spit and a small wash-over occurred just beyond the northernmost house. In spite of the waves having been somewhat smaller during the 1976 storm, the dune retreat was comparable in 1972 and 1976, so it appears that the higher tide levels of 1976 helped make up for the lower wave energies.

Tides can be augmented by storm surge if winds blow toward the land and pile up water against the coast (chapter 3). However, measurements of tide levels at Newport during the 1972 and 1976 storms revealed little increase in water levels due to storm surge (Komar and McKinney 1977; McKinney 1977). This is understandable in that the erosion during December 1972 was produced by a distant storm, and the winds blowing at the coast itself were weak. The storm of 1976 passed quickly over the coast as a front and apparently did not have sufficient time to pile up water along the beaches. This does not preclude the possibility that storm surge could be an important factor in future erosion if circumstances are right for its generation.

Mean water levels along the Northwest coast are highest during the winter months, and seasonally high water levels undoubtedly have played a role in the erosion of Siletz Spit and other areas of the coast. Their contribution was perhaps most evident during the El Niño winter of 1982–83 and the erosion of Alsea Spit (see chapter 7).

Our investigations of the erosion on Siletz Spit illustrate that several ocean processes are important factors in beach erosion. The stage is set during the fall and early winter by the general retreat of the beach under higher wave energy conditions, and especially by the development of rip currents that cut embayments deep into the beach. The actual erosion and property losses occur when a major storm strikes the coast and the rip embayments focus the wave attack; property losses under such conditions are largely limited to embayment areas. The severity of the erosion depends primarily on the energy and size of the storm waves but is also influenced by factors that determine water levels, such as tides. Although this model for erosional processes was developed primarily on the basis of observations at Siletz Spit, it is basically the same for the entire Northwest coast, with differences depending on the specifics of each beach site. For example, Siletz Spit has a coarse-sand beach, and such beaches respond more rapidly than fine-sand beaches to storm waves, and therefore undergo larger profile changes. Rip current embayments on coarse-sand beaches tend to be narrow and deep, and accordingly play a more important role in focusing the erosion than they do on fine-sand beaches. Some of these differences will become apparent when we examine erosion problems at other locations along the Northwest coast.

Although the erosion of Siletz Spit has been caused primarily by natural ocean processes, one human-induced factor can be added to the equation: beach sand mining. A sand-mining operation took place at Gleneden Beach, immediately south of the spit, between 1965 and 1971. The company's records indicate that some 110,000 cubic yards of sand were removed from the beach during that time. The mining was then halted because of its probable role in contributing to the erosion of the spit.

The significance of the sand-mining operation can be seen in its impact on the natural budget of the beach sand—the gains and losses of sand from the beach system as a whole. Siletz Spit is centered on a pocket beach, or littoral cell, that extends from Cascade Head on the north to Government Point in the south, a total along-coast length of 15 miles (fig. 6.8). Like other beaches on the Oregon coast, this littoral cell appears to be self-contained, with little if any sand movement around its bounding headlands (see chapter 3). The mineral content of the sand on the beach indicates that the Siletz River is not its primary source. Instead, it appears that much of the sand on the present-day shore has been derived from erosion of sea cliffs.

An analysis of the sediment budget of a beach requires the determination of the volumes of sand contributed to the beach by various sources (fig. 6.8). This includes only volumes of sand sufficiently coarse to remain on the beach; finer grains that immediately move offshore into deep water are excluded. We estimated the amount of sand supplied by sea cliff erosion by multiplying the rate of cliff retreat by the area of the cliff and the percentage of the eroded sediment coarse enough to remain on the beach. This calculation is only an approximation in that the rate of cliff retreat and the other factors are only roughly known. Nevertheless, we determined that the volume of sand removed by the sand-mining operation was comparable to the amount contributed by cliff erosion, the principal source of sand to the beach.

The available evidence indicates that before the sand mining, the beach was approximately in equilibrium with its sand income and losses. Aerial photographs dating back to 1939 and earlier surveys indicated that the beach had neither grown nor diminished in extent prior to the mining. According to our calculations, the quantity of sand added to the beach from sea cliff erosion was just about the volume needed to build the beach upward and maintain its elevation relative to the rising level of the sea (on the order of 10 inches per century at this location; Shih and Komar 1994).

Thus, something of a balance had been achieved. The rising sea level increased erosion on the sea cliffs, which in turn supplied more sand to increase the volume of the beach, which limited the amount of cliff erosion. This natural balance was interrupted by the sand-mining operation. The

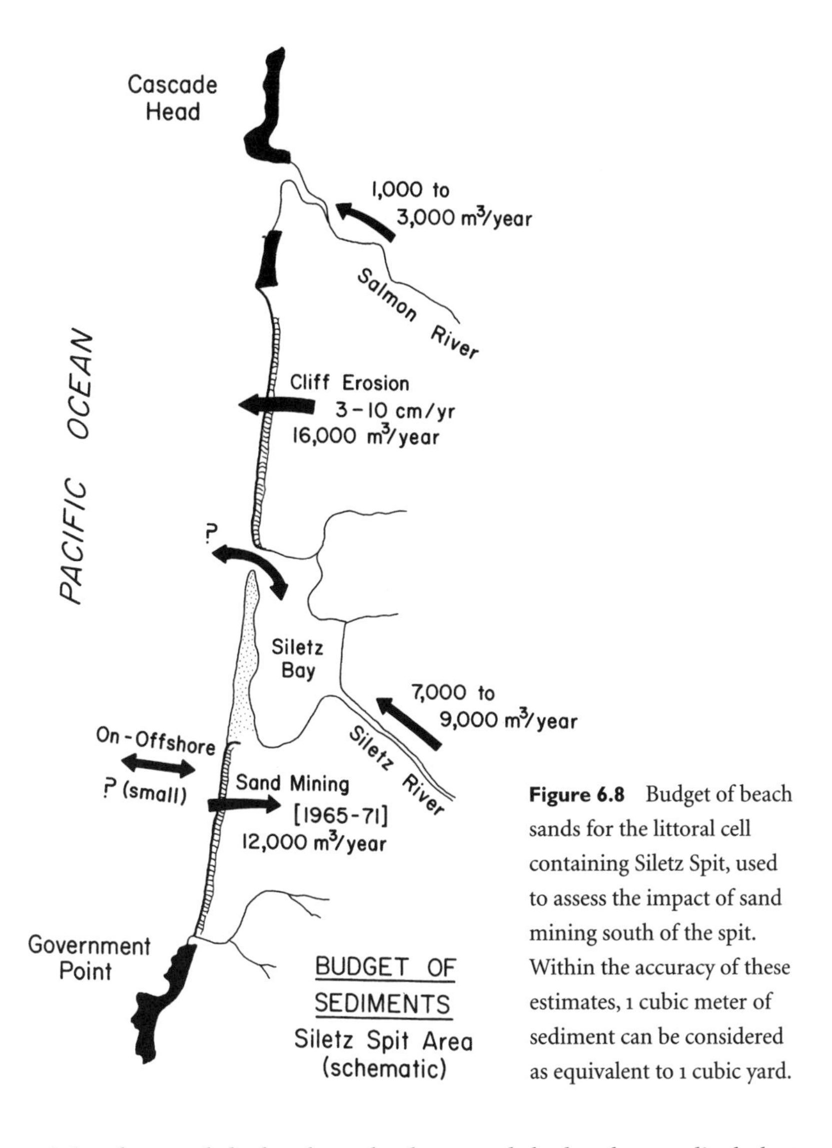

Figure 6.8 Budget of beach sands for the littoral cell containing Siletz Spit, used to assess the impact of sand mining south of the spit. Within the accuracy of these estimates, 1 cubic meter of sediment can be considered as equivalent to 1 cubic yard.

mining decreased the beach sand volume, and the beach accordingly lost some of its buffering ability. This led to accelerated erosion of the sea cliff and sand spit, at least until the additional erosion had contributed enough sand to the beach to make up the deficit caused by mining. Each cubic yard removed by the mining operation was eventually replaced by a cubic yard of sand derived from the erosion of Siletz Spit and neighboring sea cliffs. Ironically, the company doing the sand mining was also initially contracted to place rip-rap along Siletz Spit to halt the erosion. When it became apparent that the mining was a factor in the erosion, the sand removal was halted by state order.

Past Erosion on Siletz Spit

Having observed the active erosion of Siletz Spit during the winter of 1972–73, we wanted to know whether such erosion was unusual. To answer this question we examined a series of aerial photographs of the spit (Rea 1975; Komar and Rea 1976). The earliest photos date back to 1939, with reasonably frequent coverage available in more recent years (fig. 6.9). The photos revealed many past episodes of erosion on the spit, before it was developed, much like the erosion we had observed in 1972–73. During one period in the mid-1960s erosion took place along nearly the entire length of the spit, but usually it was concentrated in 100- to 600-yard segments, indicating that rip current embayments played a role in it. The sequence of aerial photos also revealed that each erosion event was followed by a long period of foredune rebuilding. After erosion cut into a section of the foredunes, drift logs washed into the area, and their crisscrossing mass trapped sand either washed there by the waves or blown inland by onshore winds (fig. 6.10). In most cases, the foredunes were naturally restored within less than 10 years after an erosion event. Thus a natural cycle existed on the spit: erosion was followed by dune reformation, which was followed by erosion.

Even before we examined the aerial photographs, however, we knew that erosion had occurred on Siletz Spit in the past. This became evident during the 1972–73 and 1976 erosion episodes, which cut back the foredunes and exposed drift logs within the body of the spit (fig. 6.11). Some of the logs had been cut with a saw, evidence of their placement since the coming of Europeans to the Northwest. Furthermore, the logs were fresh rather than rotten. Those observations combined with the aerial photographs demonstrated that many of the houses on Siletz Spit are in a zone that has undergone repeated erosion. Some of the houses even rest on the edge of an ero-

Figure 6.9 Aerial photos of Siletz Spit taken during February 1976 show rip current embayments on the ocean beach and a crescent-shaped "bite" eroded out of the bay side of the spit by the river flow.

sion scarp that formed along the length of the spit during the early 1960s, just prior to development. This scarp, now reduced to an elevation change in the dunes, is visible at the base of the houses in figure 6.2 and in the aerial photos of figure 6.9.

It is unfortunate that such an aerial photo study was not undertaken before Siletz Spit was developed. Setback lines could have been established to

Figure 6.10 Drift logs filling eroded rip current embayments on Siletz Spit help to trap blowing sands and rebuild the foredunes.

Figure 6.11 Sawed drift logs exposed within the eroding foredunes on Siletz Spit during 1976 demonstrate that this area has previously eroded and reformed.

keep homes out of the foredune area of repeated erosion, eliminating much heartache and expense. And the spit might have remained in a more natural state rather than becoming an armored peninsula.

Bay-Side Erosion

As if Siletz Spit did not have enough problems with ocean-side erosion of its beach and dunes, it has also suffered from prolonged erosion on its bay side. The bay-side erosion is caused mainly by the flow of the Siletz River, which strikes the back side of the spit just before exiting from the bay through the inlet (fig. 6.12). The erosion has cut a crescent-shaped bite from the spit that is apparent in aerial photographs (fig. 6.9). People living in the narrow section of the spit that is subject to both ocean-side and bay-side erosion are justifiably nervous (fig. 6.13).

Aerial photographs also document the bay-side erosion and the resulting decrease in the width of the spit. We determined the progress of the erosion by measuring specific points on the photographs (diagrammed in fig. 6.14; Rea 1975) and comparing them with the original section survey done in 1875. At that time the width of the spit, C, was 535 feet. By 1939, the width had decreased to 335 feet. At the time of our analysis in 1973, the width was only 171 feet. The progressive increase in the distance across Siletz Bay (line B) shows that the 364 feet of spit width lost between 1875 and 1973 was due entirely to erosion on the bay side. The edge of the ocean-side foredunes fluctuated in response to cycles of beach erosion followed by dune reformation but did not change in overall position during that century. The erosion on the bay side appears to have been fairly steady through the years, at least until the spit was developed and rip-rap was placed to limit further losses.

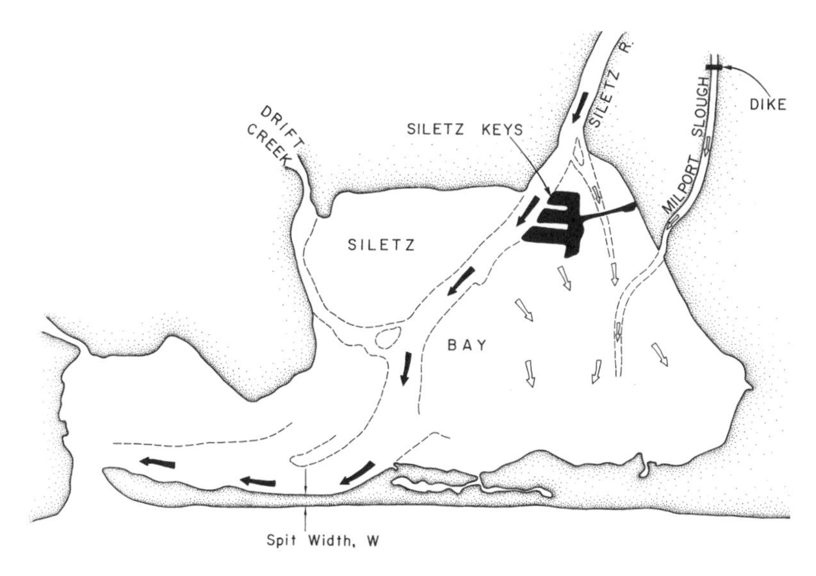

Figure 6.12 Water flow patterns in Siletz Bay are influenced by the artificial fill of the Siletz Keys development and the dike blocking Milport Slough. The fill forces more of the river flow to remain in the main river channel rather than dissipating out into the bay, resulting in increased erosion of the back side of the spit.

Figure 6.13 A house located on the narrowest portion of Siletz Spit is threatened by both ocean-side and bay-side erosion.

It is difficult to determine when the bay-side erosion began. If the rate between 1939 and the present is projected back through time, the results suggest that erosion began about the turn of the century (Rea 1975). Historic evidence such as comments by settlers regarding channel migrations and clam populations support this estimate. The erosion may have begun as a result of the natural meandering of the river flow within the bay, but logging and farming in the river's drainage basin also began at that time and may have had some influence. Such operations could have increased silt

Figure 6.14 An analysis of the decreasing width of Siletz Spit due to erosion on its bay side, using measurements obtained from old charts and aerial photographs. The spit width, C, has decreased over the years, and this corresponds to the increase in the distance B, showing that the erosion occurred mainly on the bay side of the spit. The distance D varied somewhat as a result of periodic erosion on the ocean side of the spit, but there has been no significant long-term retreat of the ocean side of the spit (1 meter ≈ 1.1 yards).

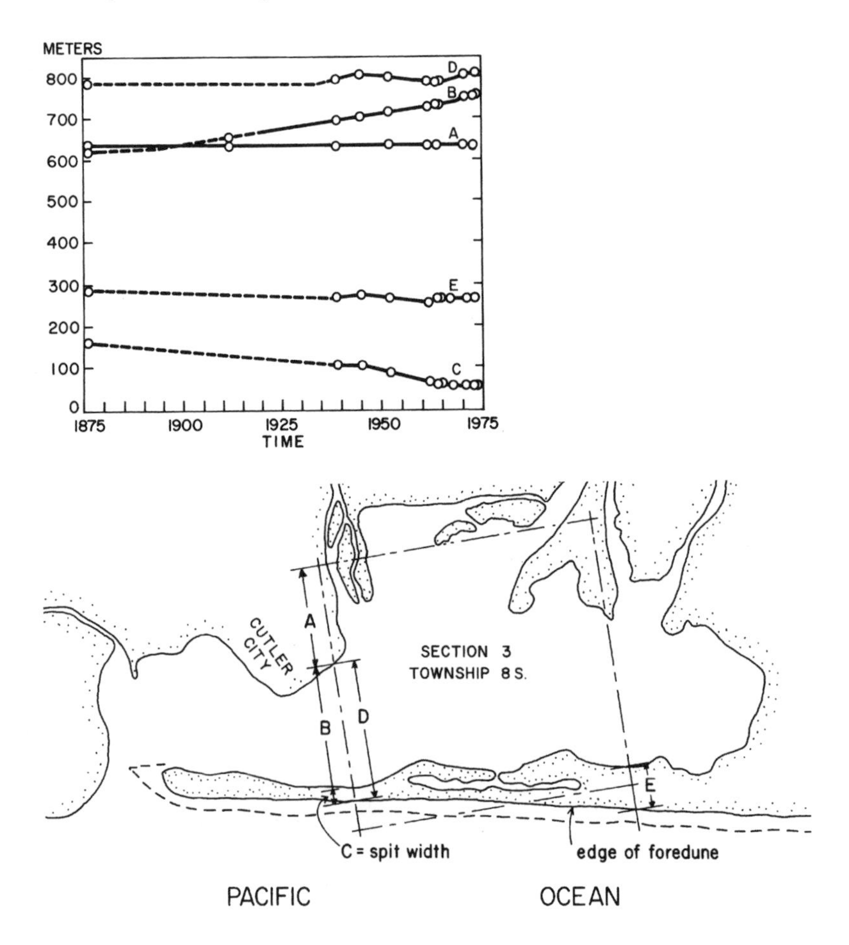

flow into the bay, and there is some indication that sediment deposition within the bay derived from Drift Creek acted to push the Siletz River over against the spit (fig. 6.12).

There is definite evidence that two recent landfills in Siletz Bay—the Siletz Keys development and a dike built across Milport Slough—have aggravated the bay-side erosion. These landfills prevent river floodwaters from spilling into the south part of Siletz Bay (fig. 6.12). A substantial volume of river water once flowed into the south bay, dissipating the energy of the floods, but the two landfills now block that flow and cause the full flood discharge of the river to be jetted against the back side of the spit. The expected increase in erosion has been reduced in part by the placement of rip-rap, but it is uncertain whether this defense will continue to be effective. During the winter of 1981–82, there was extensive erosion along the bay side of the spit as a result of exceptionally high river floods, which combined with waves generated by winds blowing across the bay from the land. Part of the access road was undermined, and extensive repairs and further additions of rip-rap were required.

In 1981, the dike across Milport Slough was replaced by a bridge, but the renewed flows have been unable to scour away the accumulated mud, so this channel is still not an effective conduit of floodwaters. Removal of the dike access to Siletz Keys (fig. 6.12) would no longer be enough to reactivate that flood-spill channel, since it is now also blocked by the approach ramp for the Highway 101 bridge over the river, rebuilt during the 1970s.

The Erosion and Breaching of Nestucca Spit

Large storm waves combined with high spring tides in February 1978 to cause extensive erosion in many areas of the Northwest coast. By that time much of Siletz Spit had been protected with rip-rap, so property losses there were minor. More critical was the erosion that occurred along Nestucca Spit on the northern Oregon coast, which threatened a new housing development and breached an uninhabited area of the spit (fig. 6.15; Komar 1978b).

The February 1978 storm was unusual in that the front hung off the coast for about four days rather than moving inland. The location of the storm immediately offshore prevented any wave spreading and energy dissipation since the fetch extended right to the shore. The Newport seismometer recorded significant wave heights of 23 feet on February 8 and 9 (fig. 6.16). The resulting erosion was intensified by the simultaneous occurrence of high perigean spring tides plus a storm surge that raised water levels by 8–9 inches above predicted tide levels. Measured high tides reached +10.2 feet MLLW, an unusually high level for the Northwest coast (see chapter 3), and substantially higher than the levels seen during the December 1972 erosion

Figure 6.15 The breach that developed on Nestucca Spit during a February 1978 storm and exceptionally high tides. From the Oregon Highway Department.

at Siletz Spit (fig. 3.7). It was the combination of a major storm with perigean high tides that produced beach erosion and the breach of Nestucca Spit.

When the storm struck, a new housing development was under construction on the foredunes at Kiwanda Beach at the north end of Nestucca Spit (fig. 6.17). The houses were protected by large quantities of rip-rap, but some empty lots were left unprotected. Dune erosion flanked the defenses and threatened homes from the side. Although erosion of the foredunes occurred along most of the spit and formed a high scarp, some areas suffered

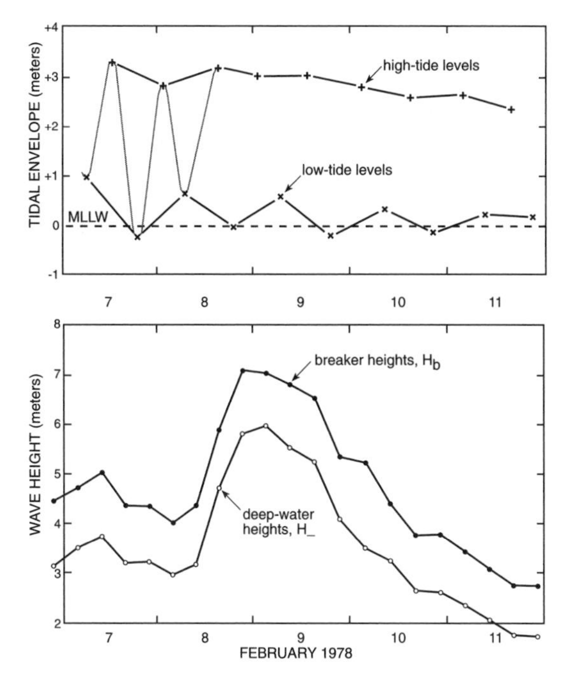

Figure 6.16 Wave and tide conditions during the February 1978 storm that breached Nestucca Spit and caused erosion of the new development at Kiwanda Beach (1 meter ≈ 1.1 yards).

greater erosion than others. As was the case on Siletz Spit, the degree of erosion was governed by the positions of rip currents and their erosion embayments. The beach at Nestucca Spit is finer grained, however, so the embayments were broader than at Siletz and the dune erosion extended over a greater length of shoreline. Again, drift logs, some of them sawed, were exposed within the eroding dunes. These logs were more rotten than those we found within Siletz Spit, suggesting that erosion episodes on Nestucca Spit are less frequent, perhaps because its beach is finer grained than the beach at Siletz (recall from chapter 3 that coarse-sand beaches respond more rapidly and to a greater degree to storm waves than do fine-sand beaches).

The most unusual aspect of the February 1978 erosion was the breaching of Nestucca Spit (figs. 6.15 and 6.18). Fortunately, the breach occurred toward the south end of the spit, well away from any homes. It developed in an area of low dunes in a narrow portion of the spit (but not the narrowest). The breach site appeared to have been determined by the location of an exceptionally large rip current and its embayment. The breach initially developed during the storm of February 8–9, but it continued to widen through March and into April, even though the waves were comparatively low. By the end of April the breach was 790 feet wide (fig. 6.18). Water washed through it only during high tides, carrying sand into the estuary to form a delta, or overwash fan.

Although breaching and wash-overs are common on spits and barrier islands along the east and Gulf coasts of the United States, where the sea level

Figure 6.17 *Top,* rip-rap was placed to protect houses that were still under construction on Nestucca Spit during the February 1978 erosion. *Bottom,* in 1988, dune sands had completely covered the rip-rap and were threatening to engulf the houses.

is rising with respect to the land (see chapter 2), the development of this breach was highly unusual for the Northwest coast. The only other spit known to have been breached during historic times was Bayocean Spit, and that breach was due to jetty construction rather than natural causes (see chapter 5). The Northwest coast is rising tectonically, and this probably accounts for the rarity of spit breaching here. It took the unusual circumstances of the February 1978 storm to produce a breach: high perigean spring tides, a significant storm surge, exceptionally energetic storm waves,

Figure 6.18 The breach on Nestucca Spit, February 1978. During high tides, the waves actively washed through the breach and into the bay.

and the development of a major rip current embayment that by chance focused the erosion on a thin section of the spit.

Nestucca Spit began to mend the following summer. Drift logs accumulated within the breach and helped to trap windblown sand, just as they had on Siletz Spit. So much sand returned to the beach fronting the housing development at Kiwanda Beach, in fact, that the rip-rap is now buried and the overabundance of sand has become a problem (fig. 6.17).

The Erosion of Cape Shoalwater, Washington

The area of greatest erosion in the Pacific Northwest, indeed the most severe along the entire West Coast, is at Cape Shoalwater on the north shore of the inlet to Willapa Bay, Washington. The erosion there has continued for approximately a century, and shoreline retreat has averaged between 100 and 130 feet per year. The problems at Cape Shoalwater are described in a 1986 paper by T. A. Terich and T. Levenseller, which also summarizes the findings of earlier unpublished studies by the Corps of Engineers.

One of the earliest navigation charts of the area was published in 1911 (fig. 6.19, *left*), but the topographic and hydrographic surveys on which it is based actually date back to 1871–91. Cape Shoalwater is shown on the map as a south-trending sand spit projecting well into the inlet connecting the ocean to Willapa Bay. At that early date, the width of the channel opening was about 3 miles. The chart on the right shows the area according to a 1911 survey. The channel width has increased to 4 miles, mainly due to the northward retreat of the Cape Shoalwater shoreline. Figure 6.20 includes a compilation of shoreline positions at various dates between 1891 and 1967, derived from aerial photographs as well as from survey charts. During that 76-year period, the shoreline retreated a total of 12,300 feet, or about 160 feet per year; the average retreat rate for the entire Cape was about 125 feet per year. The erosion has slowed somewhat since 1967 and now averages about 100 feet per year (Terich and Levenseller 1986).

A century ago, there were two channels with small shoals midway across the inlet (fig. 6.19, *left*). By 1911 (fig. 6.19, *right*), there was a single deep channel at the north of the inlet, and a large shoal had developed adjacent to Leadbetter Point, the tip of the Long Beach Peninsula. However, Leadbetter Point itself had not grown to the north, so the total width of the inlet had increased due to the erosion of Cape Shoalwater. The trends noted in these two charts—northward migration and progressive deepening of the channel within the inlet—have been the main factors in the erosion of Cape Shoalwater. This is documented more fully in the series of inlet cross sections compiled in figure 6.21. The migrating channel has impinged on the cape, resulting in its erosion. Although the shoreline retreat on the cape has been the primary concern because it has resulted in the loss of thousands of acres of land, it is apparent that the shoreline erosion is but a small part of much larger changes within the inlet.

Although Cape Shoalwater has been eroding faster and for a longer period than any other site on the Pacific coast of the United States, the prob-

Figure 6.19 *Left,* an 1871–91 Coast and Geodetic Survey chart showing the inlet to Willapa Bay with Cape Shoalwater at the north and Leadbetter Point (the Long Beach Peninsula) to the south. *Right,* a 1911 survey of the same area. From Terich and Levenseller 1986.

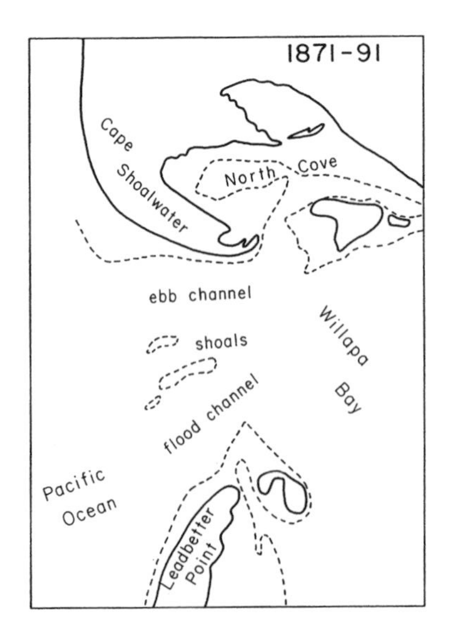

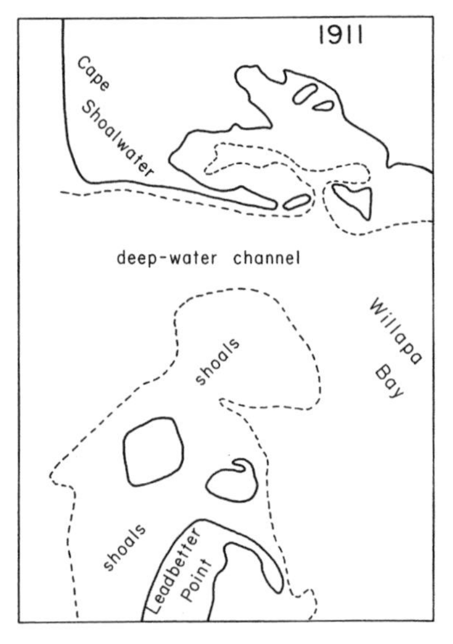

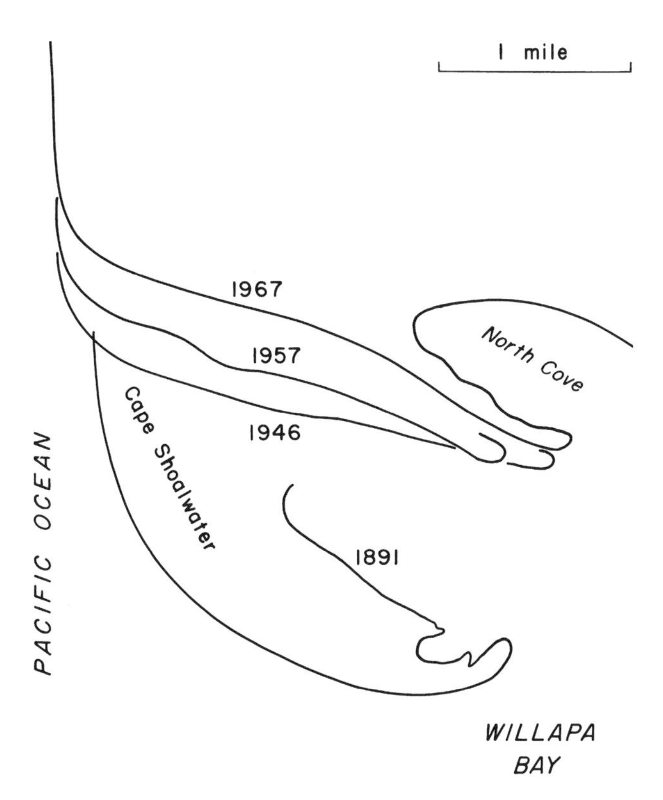

Figure 6.20 Compilation of shorelines along Cape Shoalwater derived from survey charts and aerial photographs. After Terich and Levenseller 1986.

lem has received little attention outside the immediate region. This is because the area is rural, and the erosion has endangered or destroyed relatively few man-made structures. A lighthouse was lost some years ago, and a couple of homes have succumbed to the erosion. The main highway along the coast was threatened during the 1970s and had to be moved inland. A historic pioneer cemetery was relocated at the same time. Retreat inland has been the main response to the erosion. During the 1960s and 1970s, the Corps of Engineers undertook lengthy studies to determine the cause of the erosion and what might be done to halt it. The Corps concluded that no short-term measures were feasible and that long-term solutions such as jetty construction would be extremely expensive, and recommended that funds dedicated to alleviating the erosion be spent instead to purchase the threatened land.

Although the shifting of tidal inlets like that at Willapa Bay is a common cause of major erosion along the east and Gulf coasts in the United States, it has not been a significant problem in the Pacific Northwest—except at Cape Shoalwater. This is due largely to the construction of jetties—struc-

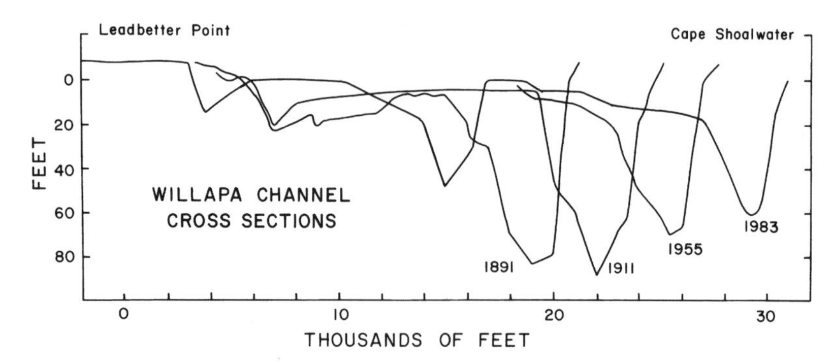

Figure 6.21 Cross sections of the inlet to Willapa Bay illustrate the progressive northward shift of the main channel; the channel cut away at Cape Shoalwater and caused shoreline erosion. After Terich and Levenseller 1986.

tures that are designed to prevent longshore channel migrations—at most inlets. There is only one recent example of channel migration causing erosion along the Northwest coast. It occurred at the Alsea Bay inlet during the 1982–83 El Niño and is described in the next chapter.

Summary

Beach erosion occurs when there is a critical combination of unusually high storm waves, rip currents that cut back portions of the beach, and elevated water levels produced by tides, wave setup, and storm surge. Generally, erosion becomes sufficiently severe to result in major property losses only when these factors act in concert. These processes were active for eons before humans decided to build homes on Siletz and Nestucca Spits and settle in the Cape Shoalwater area. Only with development did the natural processes of erosion become a "problem."

7 The 1982–1983 El Niño—
An Extraordinary Erosion Event

A decade ago, El Niño was thought to be no more than a shift in currents and a warming of the ocean waters west of South America, of interest only because it caused extensive fish kills off the coast of Peru. No one imagined that El Niños have wide-ranging consequences, including playing a major role in beach erosion along the west coast of the United States. Awareness of that came with the 1982–83 El Niño, an extraordinary event that was connected with erosion of unusual magnitude along the shores of California and Oregon. Most of the media attention was focused on the damage done to expensive real estate in southern California, particularly on the loss of homes in Malibu Beach owned by movie stars. But beach erosion and property losses were also widespread on the coast of Oregon.

The natural processes usually involved in beach erosion, those discussed in chapters 3 and 6, were also active during the 1982–83 El Niño, but at much greater intensities than normal. In addition, there were unusual events that enhanced the overall erosion problems and caused them to continue well beyond 1982–83. Erosion induced by that El Niño has given us a new appreciation of ocean waves and currents and their capacity to attack the beaches of the Northwest.

El Niño as an Atmospheric and Oceanic Phenomenon

Upwelling that brings deep ocean waters to the surface is a normal occurrence off the coast of Peru, as it is off the Oregon and Washington coasts. The cold water, which is very high in nutrients, has made the Peruvian fisheries among the richest in the world. Every few years, however, the system breaks down and the water becomes warm; fish, sea birds, and other marine life die en masse. The event usually develops during the Christmas season—hence its name, El Niño, or "The Child."

It once was thought that the onset of El Niño off Peru was caused by the cessation of local coastal winds that produced upwelling. This view changed when the physical oceanographer Klaus Wyrtki demonstrated that the local winds do not necessarily diminish during El Niños (Wyrtki 1975). Wyrtki showed that El Niño is in fact triggered by the breakdown of the equatorial trade winds in the central and western Pacific (fig. 7.1), far from the Peruvian coastal waters where its chief impact is felt. Wyrtki concluded that during normal periods of strong southeast trades, there is an elevated sea level in the western equatorial Pacific with an overall east-to-west upward slope of the water surface along the equator. You can duplicate the effect by blowing steadily across a cup of coffee; the surface of the coffee becomes highest on the side away from you. If you stop blowing, the coffee surges back and runs up your side of the cup. The process is similar in the ocean when the trade winds stop blowing during El Niño. The potential energy of the sloping water surface is released, and it is this release that produces the eastward flow of warm water along the equator toward the coast of Peru, where it kills fish not adapted to survive in warm water.

Associated with the eastward movement of warm water along the equator is a wavelike bulge in sea level. The Coriolis force, which results from the rotation of the earth on its axis, causes currents to turn to the right in the Northern Hemisphere and to the left in the Southern Hemisphere. Since the water released during El Niño flows predominantly eastward along the equator, the Coriolis force acts to confine the wave to the equatorial zone, constantly turning it in toward the equator. This prevents the sea level bulge from dissipating by turning to the north and south away from the equator. The eastward progress of the sea level wave can be monitored at tide gauges on islands near the equator. Wyrtki first demonstrated the existence of the eastward-moving wave during the 1972 El Niño by comparing the tide records at several of those islands (fig. 7.1; Wyrtki 1977). Figure 7.2 shows the

Figure 7.1 The equatorial Pacific currents and westward trade winds. Tide gauges on the islands shown here can be used to follow sea level "waves" associated with El Niño.

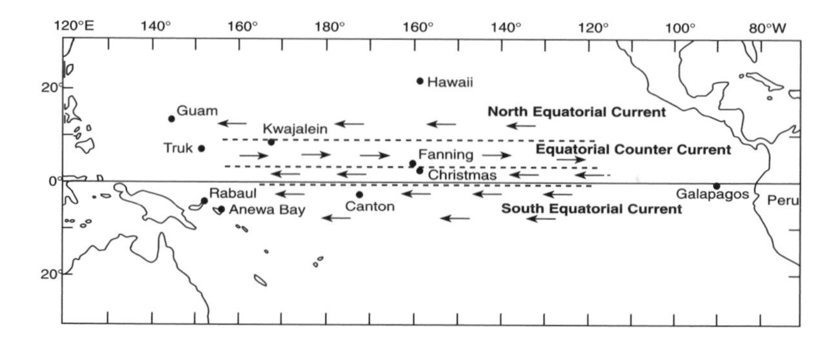

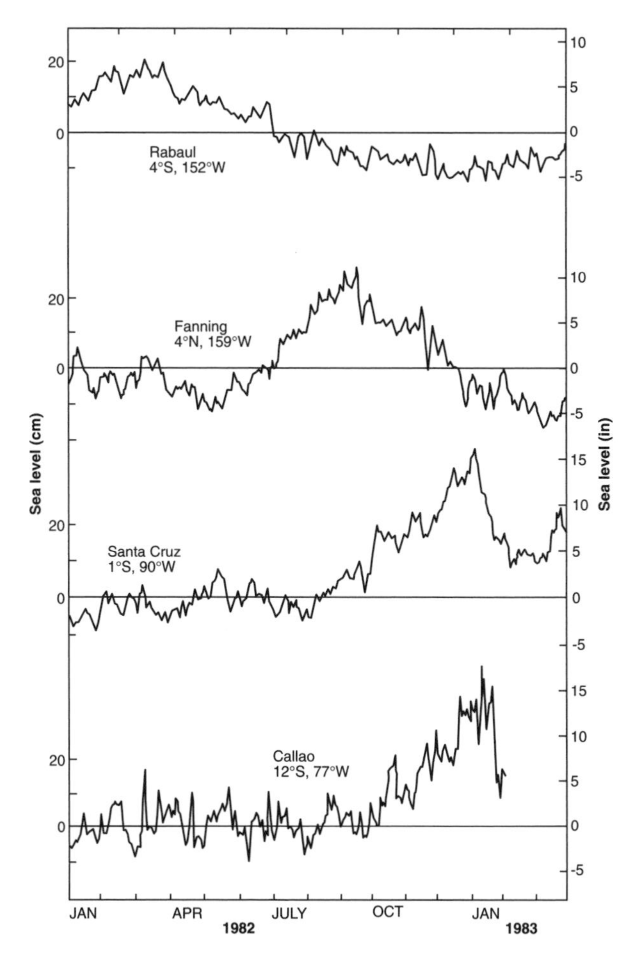

Figure 7.2 The sea level wave during the 1982–83 El Niño measured at a sequence of islands from west to east near the equator (fig. 7.1), and finally at Callao on the coast of Peru (10 centimeters ≈4 inches). From Wyrtki 1984.

similar progress of the sea level wave during the 1982–83 El Niño (Wyrtki 1984). The tide gauge records the figure depicts show the passage of the released sea level wave as it traveled eastward across the Pacific, affecting in sequence Rabaul, Fanning, Santa Cruz, and Callao Islands. Sea level at Rabaul in the western Pacific reached a peak in February 1982 and then began to drop. The crest appears to have passed Fanning Island south of Hawaii in late August, Santa Cruz in the Galápagos at the end of the year, and to have reached Callao on the coast of Peru in January 1983.

The water level changes associated with these sea level waves are quite large (fig. 7.2). They typically involve variations up to 20 inches and take place within a relatively short period, 4–6 months. In contrast, the global rise in sea level caused by the melting of glaciers (see chapter 2) has amounted to only 4–8 inches during the past century.

When it arrives at the coast of South America, the wave splits, and the two separated parts respectively move north and south along the coast.

Now the wave is held together by the inclination of the continental shelf and slope, and by the combined effects of wave refraction over the slope and the Coriolis force, which again prevent the bulge from flowing out to sea and disappearing. The wave may travel as far north as Alaska (Enfield and Allen 1980). As it moves northward, at a rate of about 50 miles a day, it loses relatively little height. The Coriolis force increases in strength at higher latitudes, so the wave hugs the coast more tightly and thereby maintains its height, even though it may lose some of its energy. The water level changes associated with such shelf-trapped sea level waves are an important factor in beach erosion along the west coast of North America during El Niños.

In summary, El Niño generates large sea level variations which take the form of a wave that first moves eastward along the equator and then splits into north- and south-moving waves when it reaches the eastern shore of the Pacific Ocean. These basinwide responses involve several months of wave travel, and at any given coastal site the sea level wave may significantly raise the water level for several months.

Figure 7.3 shows the monthly mean sea levels measured by the tide gauge in Yaquina Bay, Oregon, during the 1982–83 El Niño (Huyer et al. 1983; Komar 1986). The maximum sea level, reached during February 1983, was nearly 60 centimeters (24 inches) higher than the mean water surface in May 1982. The thin solid line in the figure follows the 10-year means for the

Figure 7.3 Monthly average sea levels measured at the tide gauge in Yaquina Bay, Oregon. Water levels during the 1982–83 El Niño (dots) exceeded all previous records (mean values are given by the solid line, the previous maxima and minima by the dashed lines; 10 centimeters ≈ 4 inches). From Huyer et al. 1983; and Komar 1986.

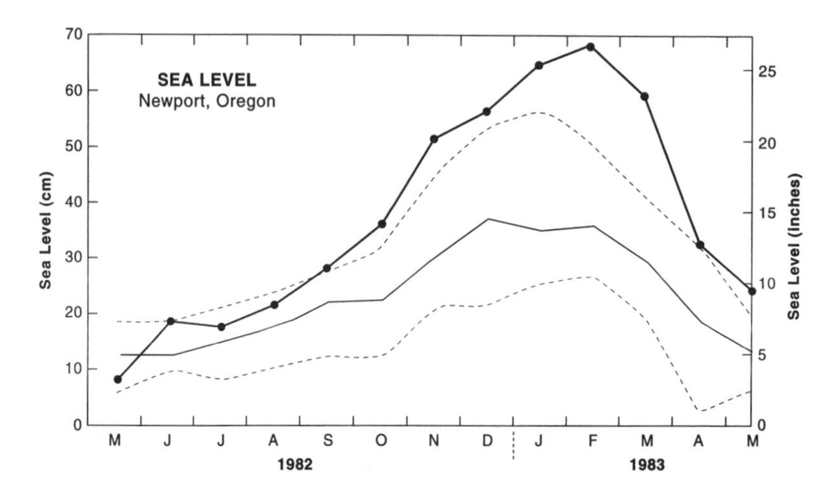

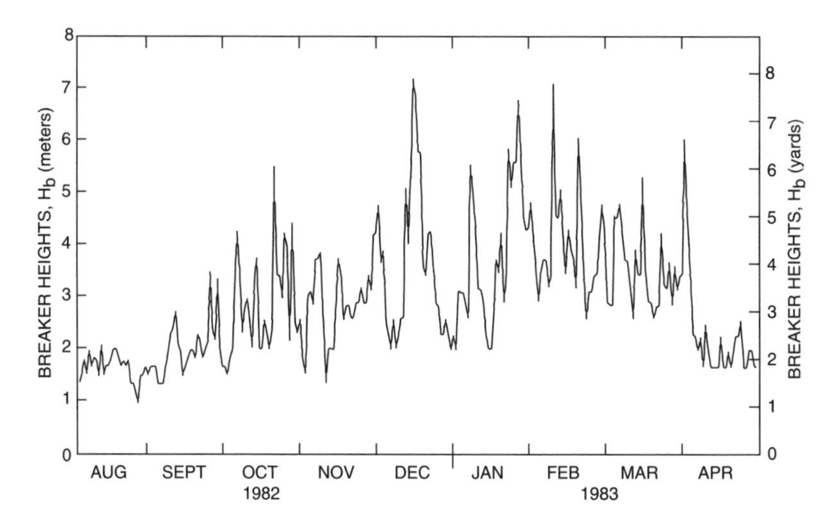

Figure 7.4 Wave breaker heights derived from the seismometer system at Newport, Oregon, during the 1982–83 El Niño (1 meter ≈ 1.1 yards). From Komar 1986.

seasonal variations, and the dashed lines give the previous maxima and minima measured at Newport. As discussed in chapter 3, these curves in part reflect the normal seasonal cycle of sea levels produced by parallel variations in atmospheric pressures and water temperatures. However, it is apparent that the 1982–83 sea levels were exceptional, reaching some 10–20 centimeters (4–8 inches) higher than previous maxima, about 35 centimeters (14 inches) above the average winter level. Most of this unusually high sea level can be attributed to the effects of a coastal sea level wave arriving from the equator. Similar extreme sea levels occurred along the coast of California at the same time and were important causes of the erosion at Malibu Beach and other coastal communities (Flick and Cayan 1985).

Wave conditions on the Oregon coast were also exceptional during the 1982–83 El Niño (Komar 1986). Measurements from the seismometer at Newport (fig. 7.4) collected daily from August 1982 through April 1983 show that several storms generated high-energy waves, with three achieving breaker heights on the order of 6–7 meters (20–25 feet). Breaker heights of this magnitude are rare in the Northwest; they occur on average only about every two years. It is therefore not surprising that extensive erosion took place during the winter of 1982–83.

The unusual severity of the 1982–83 storms illustrates the fact that there is more to El Niño than a decline in the trade winds at the equator and fish kills off Peru. In fact, as residents of the Northwest learned firsthand, El Niño represents a major disturbance of meteorological and oceanographic conditions throughout the Pacific. Also associated with the 1982–83 El Niño were droughts in Australia and floods in the United States and in the Peruvian desert. The high-altitude jet streams narrowed and intensified, and spun off cyclonic storms over the Pacific that were stronger than usual. The jet streams were also farther south than normal, so the storms crossed the North American coast in southern California rather than passing over the Northwest, giving places like Malibu Beach a taste of wave energies to which they are not accustomed.

The erosion that occurred on the Oregon coast during the 1982–83 El Niño was in response to these combined processes. The large storm waves that struck the coast arrived at the same time the sea level was approaching its maximum (figs. 7.3 and 7.4). High spring tides were also a factor. During the December 1982 storm, high tides reached +11.0 feet MLLW, 23 inches higher than the predicted tide, due to the raised mean sea level. The tides during the January 1983 storm were still more impressive—+12.4 feet, 34 inches higher than predicted. This pattern continued during the February 1983 storm, when high tides up to +10.3 feet—17 inches above the predicted level—were measured. These tides were exceptional for the coast of Oregon, where a spring-tide level of +9.0 feet is normal (chapter 3).

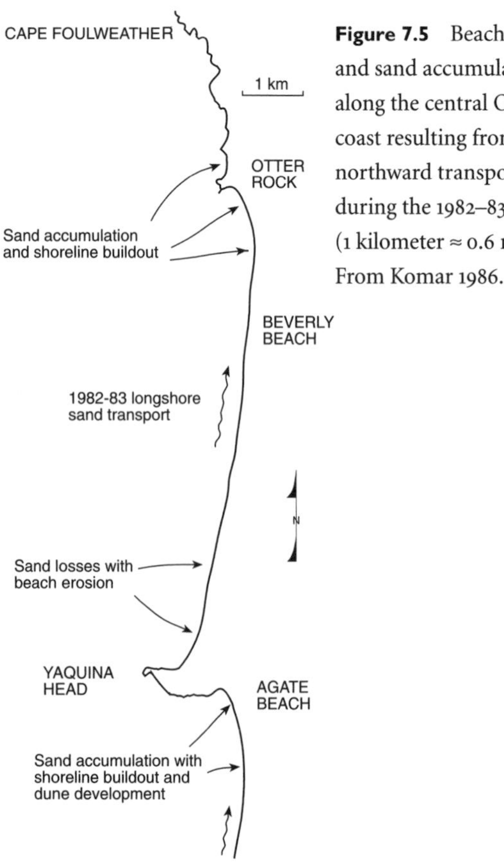

Figure 7.5 Beach erosion and sand accumulation along the central Oregon coast resulting from the northward transport of sand during the 1982–83 El Niño (1 kilometer ≈ 0.6 mile). From Komar 1986.

Responses of Oregon Beaches to El Niño

As expected, the intense storm activity and high water levels during the winter of 1982–83 cut back the beaches of the Northwest coast. However, for a time the patterns of erosion were puzzling. There were numerous reports of erosion along the coast, yet beaches in some areas clearly were building out. It took some time to determine what was happening.

Normally, summer waves approach the coast from the northwest and winter waves arrive from the southwest, so there is a seasonal reversal in the direction of sand transport along the beaches. As discussed in chapter 3, most of the Oregon coast consists of a series of large littoral cells, pocket beaches separated by headlands. Over the long term there is something of an equilibrium between the north and south sand movements within any pocket, yielding a net littoral drift of zero.

This equilibrium condition was upset during the 1982–83 El Niño as a result of the southward displacement of the storm systems. The waves approached the Oregon coast from a more southwesterly direction, and this together with the high wave energies of the storms caused an unusually large northward movement of sand within the littoral cells (fig. 7.5). The result was sand erosion at the south end of each pocket beach and sand accumulation at the north end. In other words, the pocket beaches reoriented themselves to face the waves arriving from the southwest, and each headland acted like a jetty, blocking sand flow south of it and causing erosion to its immediate north. Figure 7.5 illustrates this pattern for the littoral cell between Cape Foulweather and Yaquina Head. Directly north of Yaquina Head the beach eroded down to bedrock, while south of the headland, at Agate Beach, so much sand accumulated that it formed a large dune field (fig. 7.6). Residents of areas north of the headlands, at the south ends of the pocket beaches, experienced some of the greatest beach and property losses that occurred during the 1982–83 El Niño. The beaches eroded far more than during normal winters, with the sand moving not only offshore into bars but also northward along the shore. Once the fronting beaches lost their sand buffer, properties north of headlands were exposed to direct attack by storm waves, often suffering further large erosional losses.

Alsea Spit Erosion

The area that experienced the greatest erosion during the 1982–83 El Niño was Alsea Spit on the central Oregon coast (fig. 7.7). Erosion there was mainly in response to the northward longshore movement of beach sand, which deflected the inlet leading into Alsea Bay (Komar 1986). Although the problem originated during the 1982–83 El Niño, the erosion continued for several years as a result of the disrupted inlet.

Figure 7.6 Sand level changes north and south of Yaquina Head (fig. 7.5). The beach north of the headland (*top*) was totally depleted of sand while large quantities of sand accumulated to the south at Agate Beach (*bottom*).

A 1978 photograph of Alsea Spit (fig. 7.7) shows the spit in the early stages of development; streets had been installed, but large-scale home construction had not yet begun. The photo illustrates the normal configuration of the spit and inlet, with a narrow mouth to the far south pushed against a rocky portion of the mainland. Alsea Spit appeared to have been relatively stable over the preceding years (Stembridge 1975b), and there was little noticeable change in its morphology before the 1982–83 El Niño.

Figure 7.7 Aerial view of Alsea Spit in 1978 showing the normal configuration of the inlet. From Komar 1986.

Ordinarily, the channel from Alsea Bay continues directly seaward beyond the inlet mouth (fig. 7.7). During the 1982–83 El Niño, however, the channel was deflected well to the north (fig. 7.8). The inlet mouth itself moved very little, the deflection instead taking place in the shallow offshore. Apparent in this aerial photograph is an underwater bar extending from the south and covered with breaking waves. It was the northward growth of this bar, which occurred as a result of the northward sand trans-

Figure 7.8 The channel leading into Alsea Bay was deflected when the longshore bar grew northward in response to storm waves that arrived from the southwest during the 1982–83 El Niño. From Komar 1986.

Figure 7.9 House on Alsea Spit located in the area most eroded during the early stages of the 1982–83 El Niño: *top,* soon after the initiation of spit erosion, before rip-rap placement; *bottom,* later, threatened from the side because the adjacent vacant lot was not protected. From Komar 1986.

port during the 1982–83 El Niño, that diverted the channel from its normal course.

The erosion on Alsea Spit continued for about three years and was directly attributable to the northward deflection of the channel. The earliest property losses occurred during the winter of 1982–83 on the spit's ocean side well to the north of the inlet (fig. 7.9). The center of this erosion was directly landward of the point where the channel turned seaward around the end of the northward-extending offshore bar. The erosion appeared to be caused by the oversteepened beach profile leading into the deep channel and by direct wave attack; waves passing through the channel did not break over an offshore bar, and therefore retained their full energy until they broke directly against the spit. Placement of rock rip-rap initially protected homes in the eroding area, but when adjacent lots were left unprotected, erosion flanked the rip-rap and continued along the sides of the houses. The house shown in figure 7.9, flanked by wave erosion of the foredunes, was deliberately burned down before it fell into the sea.

The erosion of the spit continued for several years while the deflected channel slowly migrated southward toward its former position. By July 1985, the opening had moved well south of its most northerly position during the winter of 1982–83. As the channel shifted southward, the center of maximum erosion along the spit similarly shifted toward the south. In September 1985 there was an abrupt increase in the rate of erosion as the focus moved onto the unvegetated, low-lying tip of the spit (figs. 7.7 and 7.8). Within just a few days, this tongue extension of Alsea Spit completely eroded away. At the same time, the deep water of the offshore channel shifted landward, directly eroding the developed portion of the spit where it curves inward toward the inlet. Seven houses were threatened by this erosion, in particular one that was adjacent to an empty lot initially left unprotected (fig. 7.10).

Figure 7.10 During the later stages of erosion on Alsea Spit, after the 1982–83 El Niño, the main center of beach and property losses shifted progressively toward the inlet. The beach had completely disappeared, and erosion was cutting back the dunes where homes had been constructed. From Komar 1986.

The beach fronting Alsea Spit grew significantly during the summer of 1986, and the tongue of sand began to reform at the end of the spit. Erosion during the winter of 1986–87 was minimal, and Alsea Spit and the inlet finally returned to their normal conditions—those that had prevailed for many years prior to the 1982–83 El Niño.

The Erosion of Netarts Spit

The effects of the 1982–83 El Niño persisted much longer in the erosion of Netarts Spit, about 60 miles south of the Oregon-Washington border. That erosion was of particular concern because it affected Cape Lookout State Park, a popular recreation site (Komar et al. 1988; Komar and Good 1989a). Netarts Spit forms most of the stretch of shore between Cape Lookout to

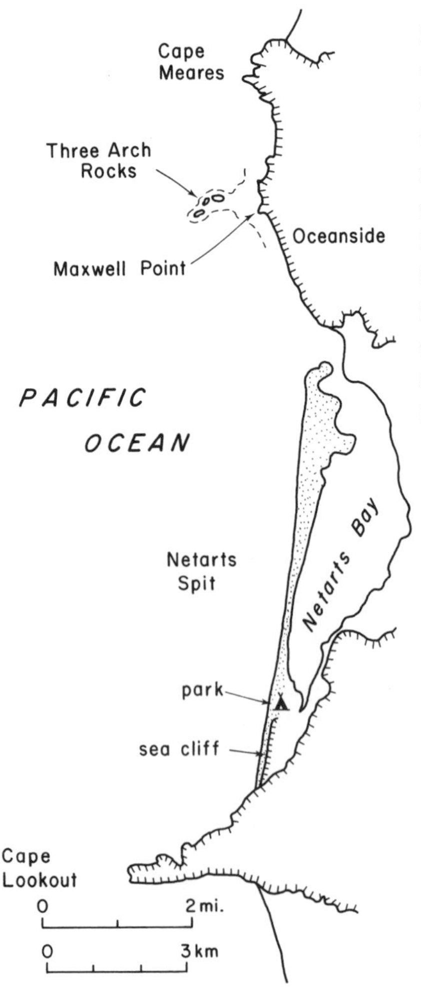

Figure 7.11 Netarts Spit on the northern Oregon coast lies within the littoral cell bounded on the north by Cape Meares and on the south by Cape Lookout. Cape Lookout State Park, where most of the El Niño–related erosion occurred, is on the south end of the spit. From Komar et al. 1988.

the south and Cape Meares to the north (fig. 7.11). It is one of the smallest pocket beach littoral cells on the Oregon coast. The spit together with its dunes contains a large volume of sand (fig. 7.12), yet there is no significant local sediment source within this isolated cell. Only a few minor streams enter Netarts Bay. Cliff erosion south of the spit and in the Oceanside area may contribute some sand to the beach, but the quantities would be small. The mineral structure of the sand forming Netarts Spit also demonstrates that most of it does not have a local source; the sand contains heavy minerals derived from rocks found in the Klamath Mountains of southern Oregon and northern California (Clemens and Komar 1988a). The presence of

Figure 7.12 Oblique aerial photo of Netarts Spit and the inlet to Netarts Bay, March 10, 1978. Photo A682-23 from the Oregon Highway Department.

Figure 7.13 The log seawall fronting the park on Netarts Spit prior to the 1982–83 El Niño. From Komar et al. 1988.

Klamath minerals in Netarts Spit can only be explained by a northward longshore sand transport during the lowered sea levels of the ice ages, when headlands did not block such movements, followed by a landward migration of the beaches when the glaciers melted and water levels rose (chapter 2). The significance of this to the recent erosion of Netarts Spit is that the spit does not have present-day sources of sand that can be added to the beach.

Before the 1980s, erosion of Netarts Spit during historic times had been minimal. In the late 1960s, a seawall was constructed at the back of the beach in the park area (fig. 7.13). Its construction was not entirely a response to wave erosion problems, but also in part to keep people from walking on the dune face, which disturbed and mobilized the dune. The sudden and dramatic erosion during the 1982–83 El Niño therefore came as a surprise. The pocket beach within the Netarts cell underwent a marked reorientation due to the approach of the storm waves from the southwest. This depleted the beach immediately north of Cape Lookout of sand, leading to erosion of the low-lying sea cliffs in that area. Of more lasting significance was that much of the sand transported northward along the beach apparently was swept through the tidal inlet into Netarts Bay, and perhaps also offshore. This effectively removed the sand from the beach system, leaving the beach depleted and less able to buffer park properties from storm erosion processes. Because of this, erosion along Netarts Spit has continued even though the direct processes of the 1982–83 El Niño stopped long ago.

Rip currents and storm waves have been the chief agents of erosion on the sand-depleted Netarts Spit. They cut back the beach in the park area, leaving much of it covered with cobbles rather than sand (fig. 7.14). The seawall was destroyed, and the erosion of park lands was substantial. Park officials considered placing rip-rap to prevent additional loss of park lands, but the fact that rip currents change position from one winter to the next, focusing erosive forces in different areas, would have made that a futile and expensive exercise.

Figure 7.14 Progressive erosion of Cape Lookout State Park following the 1982–83 El Niño: *top,* the destruction of the log seawall and the initiation of dune erosion during October 1984 (photo by J. W. Good); *bottom,* erosion during the winter of 1988 left the beach composed of cobbles and gravel rather than sand. The I-beams of the log wall are now located at mid-beach. From Komar et al. 1988.

The fundamental problem at Netarts Spit is the depleted beach, and the most recent conservation efforts have been directed toward solving that problem. State parks officials have considered beach nourishment, bringing in sand from elsewhere to replenish the beach sand, as one solution. Sand nourishment would restore the beach along its full length, making it once again both a buffer and a recreational site. The Corps of Engineers's yearly dredging operations within Tillamook Bay and the Columbia River are possible sources of sand for the project, but a more logical source would be the sandy shoals within Netarts Bay itself, in effect returning sand to the beach from which it was swept during the 1982–83 El Niño and afterward. An associated positive effect would be the restoration of the bay, which has undergone considerable shoaling. However, Netarts Bay contains many acres of protected wetlands and has the highest diversity of clam species of any estuary in Oregon. The benefits to the bay from dredging and sand removal would have to be balanced against the probable negative impacts of such operations.

El Niño and Previous Erosion

The importance of the 1982–83 El Niño to erosion along the Northwest coast raises the question of whether previous El Niños played a similar role. Although not truly periodic, El Niños occur some 20 to 25 times per century. Dr. William Quinn of Oregon State University investigated the historic occurrences and classified them according to their intensity. Table 7.1 lists El Niños since 1900 that have been assessed as "moderate" or "strong." Recall that the two major erosion periods on Siletz Spit took place during the winters of 1972–73 and 1976 (chapter 6)—both El Niño years. On the other hand, the serious erosion that took place during 1978 and led to the breaching of Nestucca Spit did not coincide with an El Niño. Unfortunately, little

Table 7.1. El Niño Intensities during the Twentieth Century

Year(s)	Intensity	Year(s)	Intensity
1991–92	medium	1929–30	medium
1982–83	very strong	1925–26	strong
1976	medium	1918–19	strong
1972–73	strong	1914	medium
1965	medium	1911–12	strong
1957–58	strong	1905	medium
1953	medium	1902	medium
1941	strong	1899–1900	strong
1939	medium		

Source: Quinn et al. 1978.

is known about earlier erosion. We have aerial photographs of the coast dating back to 1939, but there are not enough of them to determine the exact timing of erosion events. Newspaper accounts of earlier erosion (Stembridge 1975b) do not show a strong correspondence between large-scale erosion and the El Niño years listed in table 7.1, but newspapers typically report local erosion, not coastwide problems.

In sum, it appears that erosion on the Northwest coast cannot in general be attributed to El Niños, although there is evidence that El Niños can intensify erosion by increasing storm intensities and raising sea levels. It is possible that the unusual storms that eroded the coast during 1972–73 and 1976 were related to El Niños. One study found a strong statistical correlation between large waves in southern California and El Niño periods (Seymour et al. 1985). The 1972–73 and 1976 storm systems were not displaced to the south, as the 1982–83 storms were, and instead passed over Oregon. There was thus no strong northward littoral drift of sand along the Northwest beaches during those earlier El Niños, and the coastal response differed from that during 1982–83, when sand movements played an important role in causing erosion.

Summary

The 1982–83 El Niño produced considerable erosion along the Oregon coast as well as in California. The main contributing factors were exceptionally high sea levels, storms that generated intense wave conditions, and the northward transport of sand along Oregon beaches. This northward movement of sand, caused by the southward displacement of storm paths during El Niño, was unusual for Oregon beaches, where near-zero net littoral sand transport normally prevails. The sand movement was particularly important in governing locations of beach erosion, which occurred primarily on the north sides of headlands. Particularly severe erosion occurred on Alsea Spit, where the northward sand transport deflected the inlet, and in subsequent years on Netarts Spit, which lost a good deal of its sand into Netarts Bay. It took several years for the beaches to return to normal, for the Alsea Bay channel to migrate back to its usual position, and for beach sand to move southward again within the littoral cells.

We now understand that El Niños affect far more than the fisheries of Peru; they can also have major impacts on erosion on the Northwest coast. Now when we hear news reports that another El Niño may be developing, we have a justifiable feeling of apprehension.

8 Sea Cliff Erosion and Landsliding along the Northwest Coast

Although the erosion of sand spits such as Bayocean, Siletz, and Alsea has been rapid and dramatic (see chapters 5, 6, and 7), the long-term progressive retreat of sea cliffs along the Northwest coast accounts for greater property losses and affects more citizens. This is especially true in Oregon, where many coastal communities are built on nearly level marine terraces or alluvial slopes emanating from the nearby Coast Range, the areas affected most by cliff erosion (fig. 8.1). Communities such as Lincoln City, Gleneden Beach, and Newport have experienced sea cliff retreat and the parallel problem of landsliding in the undercut bluffs. State parks are being lost as cliff erosion and landsliding destroy parking lots, picnic areas, and camping facilities. In total, cliff retreat and landsliding affect hundreds of miles of the Oregon coast.

Sea cliff erosion also occurs along the northern half of the Washington coast, but it is considered to be less of a problem there due to the light development. Only a few homes and condominiums are in the path of the erosion. In Olympic National Park, cliff retreat is accepted as a natural process rather than viewed as a problem.

Considering the extent and importance of sea cliff erosion and landsliding to many citizens and communities, it is surprising how little is known about these processes and their associated problems. We do not even know the rates of cliff retreat in most areas of the Oregon coast, and so cannot determine adequate setback distances for safe development. In this chapter we will examine what is known about the extent and processes of sea cliff erosion and landsliding along the Northwest coast.

Processes of Sea Cliff Erosion

Sea cliff erosion is often viewed as the process of waves attacking and undermining the cliff, which in turn triggers landsliding or sloughing of the

Figure 8.1 Examples of sea cliff erosion on the Oregon coast: *top,* Lincoln City; *bottom,* south of Newport.

upper portions of the undercut bluff. This view is oversimplified in that a large number of processes can be involved and the cliff may respond to them in a number of ways. Figure 8.2 summarizes the erosion processes and the factors that govern rates of cliff retreat. These include the energy of the waves and their run-up intensity, tides, and sea level—factors that determine the elevation of the water against the cliff and hence the position of the wave attack. The width of the beach fronting the cliff is clearly important, too, because it controls the degree to which the beach buffers the cliff from the attacking waves. Relevant to the buffering ability of the beach is the presence or absence of rip currents, which hollow out embayments in the beach and bring waves closer to the cliffs. An example of the resulting erosion is shown in figure 8.3, a photograph from Gleneden Beach, Oregon, where a rip current embayment directed the wave attack during high tides. In this case the erosion was limited to four or five lots and two houses, the longshore extent of the rip embayment. The processes that cause sea cliff erosion are the same as those discussed in chapters 3 and 6, although there the emphasis was on sand spit erosion. The ocean and beach processes are essentially the same in both cases, but the erosion of cliffs is, of course, much slower than that of sand spits, where the loose sand of the foredunes offers virtually no resistance to wave attack. On the other hand, cliff retreat is permanent, whereas foredunes may subsequently build back out by natural processes and restore lost property (see chapter 6).

A series of laboratory experiments conducted in Japan to simulate sea cliff erosion illustrate the role of the beach and its sediments (Sunamura 1983). Artificial cliffs were constructed of loosely cemented sand. Initially there was no fronting beach, so the cliff was under the direct attack of waves (generated by a paddle in the laboratory wave basin). As the cliff retreated, the erosion generated a supply of sand that accumulated at the base of the cliff. Initially the released sand increased the rate of cliff erosion because the waves used it as a "blasting" agent. Later, however, after more sand had accumulated and a beach had developed, the beach became a buffer that caused the waves to break offshore, away from the cliff. At that

Figure 8.2 A summary of the many processes and factors involved in the erosion of sea cliffs.

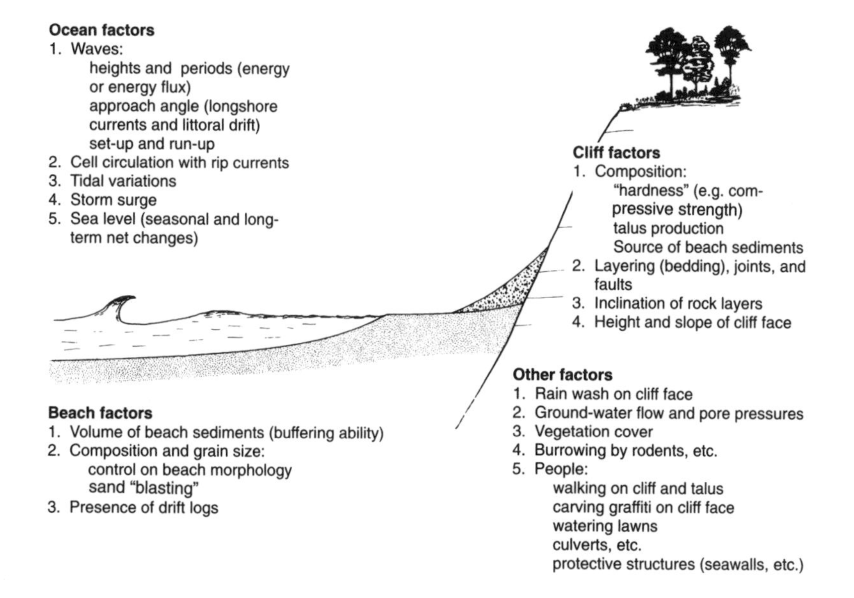

Figure 8.3 An example of rapid cliff retreat in Gleneden Beach, Oregon. A rip current cut an embayment through the beach, allowing waves to attack and undercut the cliff.

stage the cliff erosion was reduced. It is likely that similar processes are at work on cliffs backing ocean beaches. The wide summer beaches act as buffers and prevent the waves from reaching the cliffs. In the winter the beaches are cut back by storm waves and nearshore currents, and the waves can reach and attack the sea cliffs. At such times, the remaining beach sand might become an agent enhancing cliff erosion.

Drift logs are common on Northwest beaches, and it has been suggested that waves sometimes crash them against the sea cliffs like battering rams, increasing their erosive force. This may indeed be the case, although I have never seen it. Generally the logs are floated offshore during storms and episodes of beach erosion. It has also been suggested that drift logs on a beach enhance its buffering ability, and therefore prevent or limit erosion. This view might be supported by the sea cliff erosion at Taft, Oregon, that followed removal of a significant portion of the logs in 1976.

The Taft beach tends to accumulate large masses of driftwood—so much, in fact, that it hinders recreational use of the beach (fig. 8.4, *top*). To remedy the situation the state of Oregon permitted log removal during the summer of 1976. Before the driftwood was removed, the cliffs backing the beach had not eroded for many years. Soon afterward, however, during the winter of 1977–78, there was major erosion of the cliff (fig. 8.4, *bottom*).

Since that time, logs have returned to the beach and there has been no subsequent cliff erosion. The obvious conclusion is that the log removal was

Figure 8.4 Cliff erosion in Taft, Oregon, may have been caused by the removal of drift logs from the fronting beach. *Top,* the large accumulation of driftwood may have protected the sea cliff. *Bottom,* erosion began soon after the logs were removed to improve the recreational use of the beach. From Komar and Shih 1993.

a primary factor in the cliff erosion at Taft, supporting the hypothesis that masses of logs do offer some protection. On the other hand, the log removal and subsequent erosion might be simple coincidence. The storm that caused most of the erosion was severe and took place during exceptionally high water levels produced by spring tides. Erosion took place at other beaches along the Oregon coast, too, not just at Taft. In addition, a large rip current embayment cut into the beach immediately south of the Inn at Spanish Head (fig. 8.4), and this clearly contributed to the erosion of the sea cliff at Taft. Although we cannot be certain that log removal was an important factor in the cliff erosion at Taft, this episode should cause us to pause before undertaking whole-scale log removal from beaches.

Cliffs respond to erosion processes in different ways because there are many factors governing their resistance (fig. 8.2). Of primary importance is the composition of the rocks forming the cliff. The difference this can make is well illustrated on the Northwest coast. We need only think of the extremely hard basalts of the headlands, which can resist the onslaught of even major storm waves, and compare them with the weaker sandstones and mudstones that more readily give way to wave attack and form the cliffs that back stretches of beach. Even these sedimentary rocks exhibit variable resistance due to differences in rock compositions.

Many cliffs consist of Tertiary mudstones and siltstones (fig. 8.5, *top*), which can be quite resistant to wave attack. Other cliffs are composed entirely of Pleistocene terrace sands (fig. 8.5, *bottom*) that are only moderately cemented and therefore succumb easily to wave attack. The cliffs in many areas of the Northwest coast are composites of less resistant Pleistocene sandstones lying above Tertiary mudstones. This layering complicates cliff retreat. Groundwater tends to flow out at the interface of the porous sandstone and the nonporous mudstone, eroding the Pleistocene sandstone layer in a process separate from ocean processes.

The composition of the cliff material also determines whether or not eroded material accumulates as talus at the base of the cliff. If too fine grained, the loosened material is carried away by rain or groundwater flow. This is the case with mudstone if it is relatively homogeneous and does not contain resistant layers that can break off as individual blocks. The Pleistocene terrace sands are more likely to accumulate as cliff-base talus (fig. 8.6, *top*), usually soon after an episode of wave attack and erosion.

The waves first remove the talus that accumulated since the previous erosion episode, perhaps several years earlier, leaving a nearly vertical cliff face. During the week or two following the erosion there is active sloughing of the terrace sands and a rapid new talus accumulation. Such sloughing may involve minor slumps of the cliff sands—effectively a vertical drop of blocks of intact sandstone—as well as increased groundwater seepage and direct runoff of winter rain from the newly exposed cliff. As time passes and the accumulating talus protects the cliff, the latter processes diminish. In many areas, cliff retreat is due more to these processes than to direct wave attack. The main role of the ocean processes is to remove the accumulated talus sheltering the cliff, permitting renewed groundwater sapping and rain wash.

The extent of the talus and its degree of vegetation cover are evidence of how recently the area has experienced wave attack. The absence of talus at

Figure 8.5 Examples of sea cliffs: *top,* the cliff at Fogarty Creek State Park is composed mainly of Tertiary mudstones with only a thin layer of Pleistocene terrace sand at the top; *bottom,* the cliff at Lincoln City consists entirely of Pleistocene terrace sandstones.

Figure 8.6 A sea cliff at Gleneden Beach State Park, Oregon: *top,* new talus began to accumulate at the base of the cliff soon after erosion had removed the older talus covered with vegetation; *bottom,* another episode of erosion removed the newly accumulated talus, leaving a vertical cliff face.

the base of a sea cliff indicates recent erosion, but only if the cliff materials are suitable for the development of talus. Where the fronting beach is narrow, wave erosion may occur each winter so that only minor talus accumulations build up during the summer months. Such areas generally have the fastest rates of sea cliff recession. In other areas, Taft being an example (fig. 8.4), wave attack is infrequent and the talus may accumulate over several years or even decades. This permits the development of a vegetation cover, even small trees. The extent of that cover and the ages of the trees are a good indication of how long ago the last erosion took place (fig. 8.7).

In some areas vegetation grows on the bluff itself as well as on the accumulated talus. Such bluff vegetation can be important in reducing erosion in that it protects the cliff from attack by winter rains and may also resist sapping by groundwater. Burrowing by rodents weakens the cliff material and funnels the groundwater, effectively enhancing erosion. Unfortunately, people do the same sort of damage when they carve graffiti or cut tunnels into the exposed bluff (fig. 8.8). Since natural processes of sea cliff erosion produce retreat rates amounting to only a few inches per year, the human factor is far from negligible. In some places it is the most important agent in sea cliff retreat.

Cliff erosion is also affected by the geometry and structure of the cliff rocks—particularly rocks that are otherwise fairly resistant (e.g., basalt headlands and Tertiary mudstones), because the bedding and fractures provide lines of weakness. A good example of structural control of basalt erosion is the wave-cut bench near Cape Perpetua on the Oregon coast (Byrne 1963). Here, the principal set of joints is oriented in a northwest-southeast direction. Although the greatest wave energy comes from the southwest, erosion is predominantly along the northwest joint direction and has produced a series of surge channels, crevices, caves, and blowholes (fig. 8.9). In general, irregularities and small bays and inlets in the headlands are governed by joints and fault-controlled erosion or by dikes and layering within the ancient lavas.

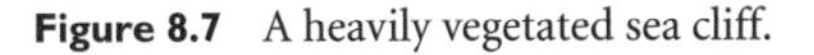

Figure 8.7 A heavily vegetated sea cliff.

Geometry and structure are also important in the erosion of Tertiary mudstones. Although faults and joints do play a role, more important here is the layering of the deposits. Mudstones were originally formed by sediments deposited on the seafloor in the deep ocean (see chapter 2), and therefore accumulated as nearly horizontal layers. Some layers were originally mud, others were silts or sandy sediments. Variations in the composition of the deposited sediments produced the layers that we now see in the cliffs backing the beaches (fig. 8.10). Some layers are more resistant than others, and so project further out from the cliff face. Particularly durable layers may break off as blocks and accumulate at the base of the cliff, where they offer some protection from erosion. Differences in the permeability of the various layers channel groundwater flow and determine patterns of sapping of the cliff face. However, the most important feature of these Tertiary

Figure 8.8 Graffiti carved into the sea cliff at Lincoln City. Such human handiwork may do more to produce cliff retreat than natural processes.

Figure 8.9 A wave surge channel on Cape Perpetua follows a northwest-oriented fracture that provided a zone of weakness for the erosion within the otherwise highly resistant basalt.

sedimentary rocks is their seaward slope (fig. 8.10). Combined with wave erosion that creates nearly vertical surfaces, such slopes yield unstable cliffs prone to massive landslides.

Landslides and Property Losses

Landslides that occur suddenly and affect large areas can have major consequences for homes, parks, and highways. It has been estimated that seaward-sloping rocks like those shown in figure 8.10 are present on more than half of the northern Oregon coast (Byrne 1964; North and Byrne 1965). The muddy consistency of their Tertiary sediments makes these cliffs particularly susceptible to landsliding, although major landslides have also occurred on the steep slopes of headlands.

The term *landslide* has been used to represent a variety of types of displacements of masses of rock, soils, and sediments. The common ingredient of all such movements is gravity, which acts on the mass, causing it to move downward and outward. The classic landslide, or slump (fig. 8.11), is the type most likely to take place on sea cliffs, which lack support in the seaward direction. Without support, stress develops in the cliff materials. When the stress exceeds the resistance of the rock, a landslide occurs. The failure often takes place along a concave up surface, and this becomes the rupture plane over which the slump moves (fig. 8.11). As the slump moves downward and out onto the beach, the upper portion of the rupture surface

Figure 8.10 A sea cliff in the Jump-Off Joe area of Newport, Oregon, shows the seaward-dipping layers within the Tertiary mudstones and the upper layer of Pleistocene marine terrace sands. The steep slope of the layers within the mudstones contributes to the massive landsliding that has occurred at this site (see fig. 8.12 and chapter 9).

is exposed, forming a nearly vertical scarp at the landward limit of the slump.

As the slump progresses, the material rotates such that originally horizontal surfaces tilt landward. This rotation is particularly evident when the slump occurs in a marine terrace. Portions of the formerly horizontal terrace now slope toward the land, and trees that once grew vertically tilt inland with uniform orientations (fig. 8.12). The farther the material moves, the more jumbled it becomes; eventually, intact portions of the terrace no longer exist and the trees slant in all directions. By the time the material flows out onto the fronting beach, it is generally a tangled mass of rocks, soil, trees, and brush. This portion comprises the *foot* and *toe* of the slump (fig. 8.11).

Waves quickly go to work to erode away this mass, which may extend across much of the beach and even into the sea. It is surprising how resistant slide masses can be to wave attack. It sometimes takes months for the waves to cut back a landslide that flowed across the beach and into the surf zone.

Wave erosion of the slump toe produces a temporary cliff, or scarp, and removes the basal support of the slide mass, creating additional instability. This type of erosion may result in repeated movements of the mass. In the case of very large slumps, the process may continue for years, with the slump mass slowly moving seaward at the same rate the waves cut away the cliff formed in its toe. Sometimes this type of movement occurs in large

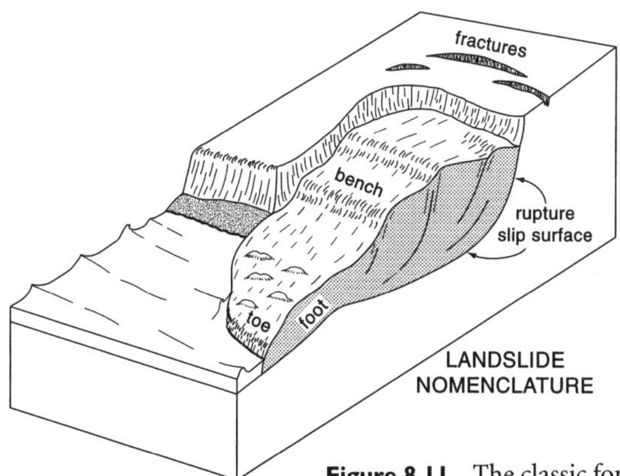

Figure 8.11 The classic form of a slump or landslide.

blocks of land that are still relatively intact, never having gone through a phase of rapid sliding. The movement may amount to only a few inches per year, but that is enough to disrupt roads and progressively shift foundations of homes. Such unstable sites are obviously undesirable locations for permanent structures, yet developers persistently build on them. A good example is the destruction of the Stratford Estates development north of Newport (fig. 8.13), where streets and sewers placed on a slow-moving landslide were quickly disrupted.

Most movement on landslides occurs during the winter months (fig. 8.14), primarily December and January, which are also the months of maximum rainfall. This is hardly surprising, since rainfall and groundwater are primary agents in landslide generation. The winter months are also a time of intensified ocean wave activity (see chapter 3), and winter waves may also contribute to landslides by undercutting cliffs and thus increasing their instability.

The largest landslides on the Northwest coast occur on headlands or within the loose debris along their immediate margins. A good example is the huge slump on Cascade Head, Oregon (fig. 8.15), which abruptly gave way in 1934 (North and Byrne 1965). The surf has cut away at the toe of the landslide, forming a high cliff in the debris. Massive landslides associated with headlands also affect developed areas. The large landslides that cross Ecola State Park on Tillamook Head, immediately north of Cannon Beach, become active every few years, disrupting access roads and other facilities (fig. 8.16; Schlicker et al. 1961; Byrne 1963).

Along-Coast Variations in Sea Cliff Erosion

One of the difficulties with managing sea cliff erosion on the Northwest coast is its extreme variability, both spatially and temporally. This variability makes it difficult to establish reasonable setback distances to ensure safe development and to judge whether or not cliff protection structures such as seawalls are justified.

My colleagues and I have focused on this variability in our investigations of sea cliff erosion along the Oregon coast, particularly that in the littoral cells on the northern half of the coast (fig. 8.17; Komar and Shih 1991, 1993; Shih 1992; Shih and Komar 1994; Shih et al. 1994; Ruggiero et al. 1996). Each cell exists as a nearly isolated pocket beach bounded on the north and south by headlands that prevent alongshore exchanges of beach sand. In addition to investigating cliff erosion within this series of cells, we also examined

Figure 8.12 Tilted trees and inclined ground on what was once the horizontal surface of a marine terrace are evidence of the disruption produced by the Jump-Off Joe landslide in Newport. These photos, taken in 1975, predate attempts to build condominiums on the landslide (see chapter 9).

Figure 8.13 A street and sewers in the Stratford Estates development north of Newport were destroyed by a slow-moving landslide. The disruption has caused the street to be "cloased."

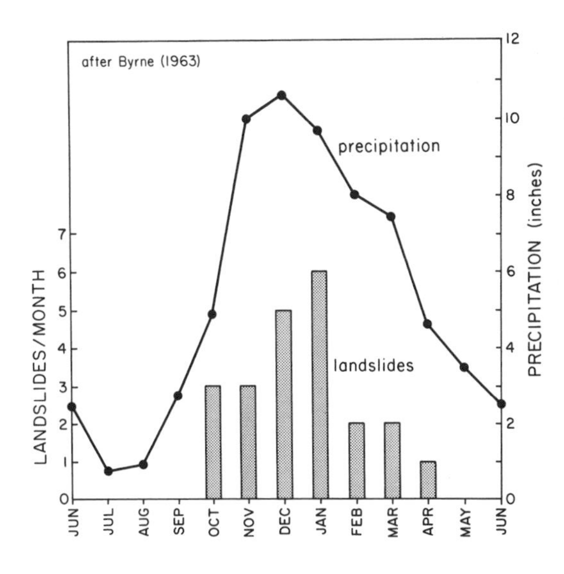

Figure 8.14 Number of landslides versus monthly precipitation amounts. Most landslides on the Oregon coast occur during the months with highest precipitation, reflecting the importance of rainfall and groundwater in the generation of landslides. From Byrne 1963.

specific sites of potential erosion along the southern half of the coast, assessing vegetation cover on the cliffs, the quantity of accumulated talus at the cliff base, and the occurrence or absence of wave attack in recent years. We identified a coastwide pattern of cliff erosion that parallels the pattern of tectonic uplift along the coast relative to the global rise in sea level (Komar and Shih 1993). Recall from chapter 2 (fig. 2.7) that the sea level along the north-central portion of the coast is presently rising at a rate faster than the rate of tectonic uplift. Sea cliff erosion is active on this portion of the coast, illustrated by the bluffs within the Lincoln City and Beverly Beach littoral cells (fig. 8.18, *middle*). The relative water level aver-

Figure 8.15 A massive landslide occurred on Cascade Head, Oregon, in 1934. Courtesy of J. V. Byrne.

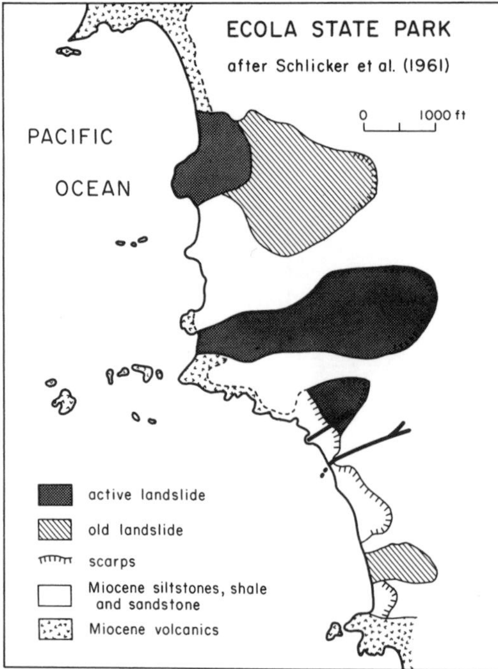

Figure 8.16 Landslides in Ecola State Park on Tillamook Head, Oregon, periodically disrupt park facilities. After Schlicker et al. 1961.

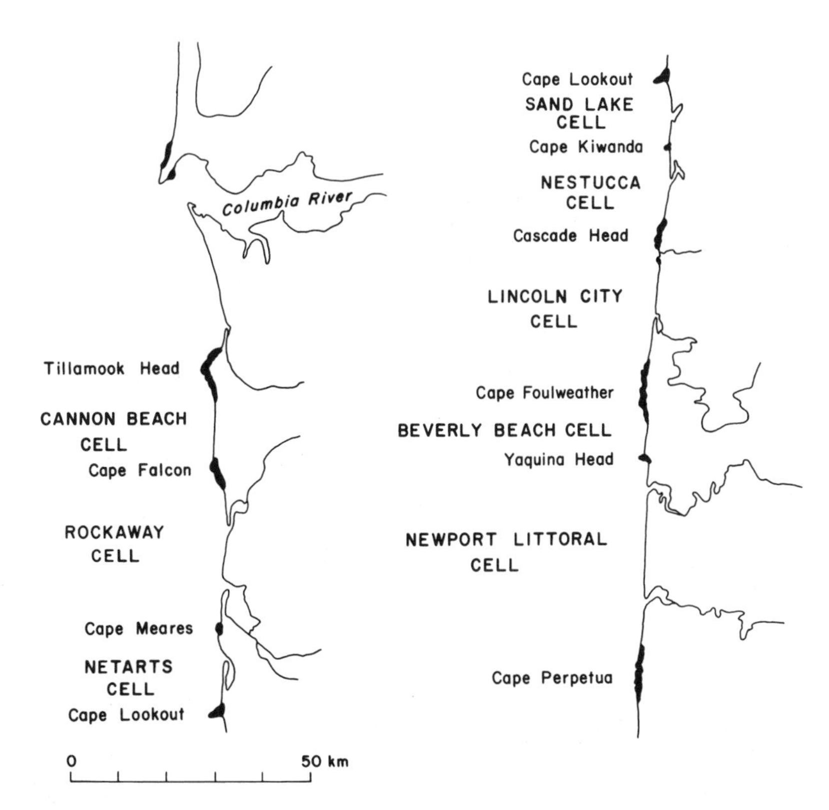

Figure 8.17 A series of littoral cells on the northern half of the Oregon coast form beach embayments between rocky headlands. Cliff erosion within each cell depends on the tectonic rise of the land relative to the increase in sea level, and on local factors such as the rock composition of the cliff and the extent of the fronting beach that buffers the cliff from wave attack (1 kilometer ≈ 0.6 mile). From Komar and Shih 1993.

ages slightly higher here each year, and this accounts for the continued cliff erosion. In contrast, there has been essentially no wave-induced cliff erosion within the Cannon Beach cell near the Oregon-Washington border in the north (fig. 8.18, *top*) or at Bandon on the south coast (fig. 8.18, *bottom*). The sea cliffs at these sites are heavily vegetated, and there has been no significant erosion resulting from direct wave attack during historic times. In addition, these sites lie within stretches of coast that are rising tectonically at rates that exceed the present rise in the global sea level (fig. 2.7). The lack of cliff erosion in these areas can be explained simply by the uplift of the land: each succeeding year the waves are less able to reach the cliffs to erode them. It is apparent, then, that the extent of cliff erosion along the Oregon coast roughly parallels the net change in relative sea level, the difference between the tectonic uplift of the land and the global rise in sea level.

Figure 8.18 Sea cliffs on the Oregon coast: *top,* in the Cannon Beach cell on the northern Oregon coast; *middle,* in the Lincoln City and Beverly Beach cells on the middle coast; *bottom,* at Bandon on the south coast. The variable rates of sea cliff erosion at these sites parallel the north-to-south trends of tectonic uplift of the coast relative to the global rise in sea level. From Komar and Shih 1993.

The minimal erosion during historic times of sea cliffs within the Cannon Beach cell and at Bandon is particularly intriguing. The little cliff retreat that occurs there is mainly the result of groundwater seepage, and not direct attack by ocean waves. Yet the steepness of the cliffs at those locations, and their alongshore uniformity without appreciable degradation by subaerial processes such as rainfall and groundwater flow (fig. 8.18, *top* and *bottom*), suggest that these cliffs experienced wave-induced erosion in the not-too-distant past. In addition to the steep cliff backing the beach at Bandon, there are a number of sea stacks in the immediate offshore, many with flat tops that continue the level of the marine terrace, further attesting to comparatively recent and major cliff erosion and retreat.

Our interpretation of both the Cannon Beach cell and the Bandon area is that cliff erosion occurred following the last major subduction earthquake, about 300 years ago (Komar et al. 1991; Komar and Shih 1993). As discussed in chapter 2, major portions of the coast may drop several feet during a subduction earthquake, which releases the strain that had previously caused coastal uplift. An abrupt coastal subsidence 300 years ago would have resulted in massive erosion and retreat of sea cliffs as well as sand spits. In the case of Cannon Beach and Bandon, the uplift that has occurred since the earthquake has been sufficiently rapid to diminish and then stop cliff erosion. The other littoral cells along the north-central part of the Oregon coast likely also experienced subsidence and massive cliff erosion 300 years ago, but subsequent uplift there relative to the rising sea level has been insufficient to completely halt cliff erosion, which is still apparent in the Lincoln City and Beverly Beach cells (fig. 8.18, *middle*).

Although cliff recession on the Oregon coast is directly related to tectonic uplift relative to sea level rise, there is a great deal of local variability (spatial and episodic) that is attributable to local factors. These include the overall ability of the fronting beach to buffer the cliffs from the waves, beach processes such as run-up and erosion within rip current embayments, and the composition of the cliff materials (fig. 8.2).

The importance of cliff composition to erosion is evident in a comparison of the Newport and Beverly Beach littoral cells with cells farther north. The sea cliffs in the Newport and Beverly Beach cells consist mainly of Tertiary mudstones. These deposits dip seaward at an angle of 30 degrees in the Nye Beach area of Newport (fig. 8.10), a factor important to the generation of the large Jump-Off Joe landslide (see chapter 9). Although also important in the Beverly Beach cell, landsliding is less catastrophic and rapid than in the Newport cell, probably because the mudstones at Beverly Beach dip seaward at lower angles. Landsliding also has been important in the Cannon Beach cell, where the cliff material consists of muddy alluvium and ancient debris. Major efforts have been made to control these slides (drainage, etc.) because they affect Highway 101. Landslides have not been a significant problem in the other littoral cells. The cliff in the Lincoln City cell is composed entirely of Pleistocene terrace sands (fig. 8.5), which generally fail in small-scale vertical falls rather than in the massive landslides typical of the mudstones.

Cliff erosion has been most active in the Beverly Beach littoral cell (fig. 8.18, *middle*). Some of the erosion is occurring in intact land masses that are slowly moving toward the sea. Further, the beach within this cell is a poor buffer because of its small sand volume. My colleagues and I documented the lack of buffering protection by measuring wave run-up elevations on the beach (Shih 1992; Shih et al. 1994). Our objective was to investigate the frequency with which waves reach the talus at the base of the sea cliff and the intensity of the swash run-up. We found that the wave swash commonly reaches the cliff base in the Beverly Beach cell but rarely does so in the other cells. This is because the beach profiles within the Beverly Beach cell have low elevations compared with mean sea level and high-tide elevations.

Most of our research has centered on the Lincoln City cell (fig. 8.19; Shih 1992; Shih and Komar 1994), which is of particular interest because of the extensive development on the cliff edge along this stretch of coast (fig. 8.1, *top*). In addition, one unusual feature of this cell enhances its scientific interest: there is a marked longshore variation in the coarseness of the beach sand, and this produces longshore changes in the beach morphology and the nearshore processes that are important to cliff erosion. The beaches on the central to southern part of the cell, including the beaches fronting Siletz Spit and the community of Gleneden Beach, have the coarsest sand (fig. 8.19). Sand grain size decreases somewhat south of there but much more so in the northern part of the cell, where the sand is finest in the Roads End area of Lincoln City.

The effect of sand size on the beach morphology is significant. The coarse-grained beach at Gleneden Beach is much steeper than the beach at Roads End, which has a very low slope. Recall from chapter 3 that coarse-sand beaches respond faster to storm waves and exhibit larger profile changes than fine-grained beaches. The large profile shifts at Gleneden Beach make that beach a poor buffer, and as a result, cliff erosion is much more active there than it is north of Lincoln City, where the cliffs are fronted by a fine-sand beach with a low slope. Rip current embayments also cut more deeply into coarse-sand beaches, and they have been a significant factor in the sea cliff erosion at Gleneden Beach (fig. 8.3). The rip current embayments on the fine-sand beach north of Lincoln City are broader in longshore extent but do not cut as deeply into the beach.

In contrast, bluff retreat in north Lincoln City is caused mainly by rainfall beating against the cliff face and groundwater seepage, aided considerably by the carvers of graffiti (fig. 8.8). The fallen material accumulates as

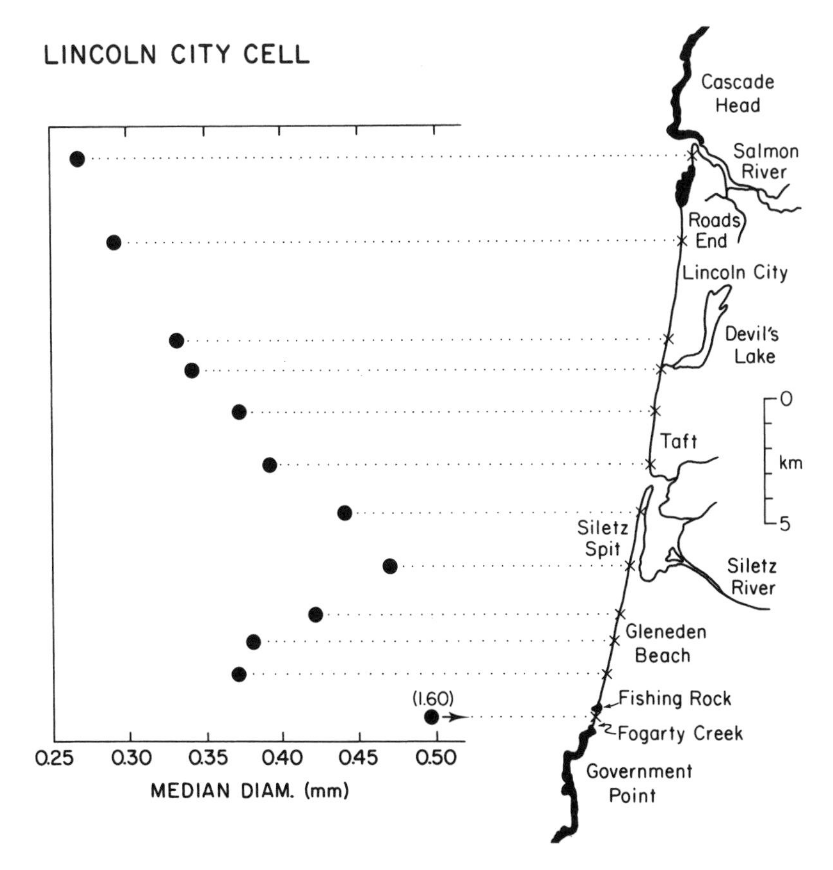

Figure 8.19 Sand grain diameters on the beaches of the Lincoln City littoral cell. The beach sand is coarsest in the vicinity of Gleneden Beach and Siletz Spit, and becomes progressively finer toward the north and, to a lesser extent, toward the south (1 inch ≈ 25 millimeters; 1 kilometer ≈ 0.6 mile). From Komar and Shih 1993.

talus, sometimes for years or decades, until it is removed by wave action during an unusually severe storm accompanied by extreme high tides.

Rates of Sea Cliff Recession

The spatial variability and episodic nature of sea cliff erosion along the Northwest coast make it extremely difficult to obtain accurate and meaningful measurements of long-term average cliff recession rates. We tried to use sequences of aerial photographs to measure these rates, but our attempts were not particularly successful, even in the Lincoln City and Beverly Beach cells, which are known to have experienced wave attack in recent years (Shih 1992; Komar and Shih 1993).

We concentrated on the Taft area of the Lincoln City cell, which eroded extensively during the winter of 1977–78 (fig. 8.4). A series of aerial photographs of that area dating back to 1939 is available. Even in the area that was thought to represent significant cliff recession, however, we were unable to determine the long-term erosion rate. The photographs show episodes of talus removal by waves followed by decades of accumulation, but recession of the top of the bluff has been too small to measure accurately on the aerial photographs. Ground photos taken over the years substantiate the negligible cliff retreat in Taft. Old photos, undated but known to have been taken in the 1920s (fig. 8.20), differ little from modern ones, confirming that there has been little retreat of the bluff top. My own photographs of the coast span some 25 years and further verify that cliff retreat has been very slow. Even in areas of the coast perceived to be undergoing significant cliff erosion, the long-term retreat rates are a couple of inches per year at most.

In a few locations the retreat has been dramatic when rip current embayments reached the base of the cliff and allowed direct wave attack, at least for a few days. That erosion is measurable on aerial photographs, but it is local and episodic, and in the long term represents a small rate of average recession.

Another factor that makes it difficult to measure cliff recession along the Oregon coast is the mass movement of the cliff itself. In the Beverly Beach and Newport littoral cells, large blocks of the mudstones, some several acres in extent, move seaward a few inches each year. These blocks remain largely intact because of their slow movement. As a block slides toward the ocean, wave erosion cuts back its seaward edge at just about the same rate as its movement. Although bluff erosion is occurring locally on these moving blocks, and homes built atop them slowly shift toward the cliff edge, the cliff itself remains approximately stationary in position as viewed in aerial photographs.

We had hoped to determine long-term recession rates that would enable coastal communities to establish setback distances to protect new construction. Although we were unsuccessful, we did learn that long-term retreat is much less than initially thought. The establishment of setback lines is a valid management approach, but it should be based on an assessment of the capacity of a particular beach to serve as a buffer and the susceptibility of the cliff material to landsliding. These factors, and thus reasonable setback distances, differ from one littoral cell to another, and secondarily within the littoral cell depending on local conditions.

Structures that Prevent Cliff Erosion

There have been many attempts to limit or prevent sea cliff erosion by constructing protective structures. These structures have taken a variety of

Figure 8.20 Photos of the Taft area of Lincoln City taken in the 1920s (*top*) and recently (*bottom*) demonstrate that bluff retreat there has been minor over the past 60 to 70 years. From Komar and Shih 1993.

forms—including rip-rap and seawalls—depending on the magnitude of the erosion and the height of the sea cliff. Some have been successful; others were a complete waste of money. Some structures have minimal visual impact; others are noted mainly for their ugliness.

The minimum protective structure is represented by a line of rip-rap that protects the toe of the accumulating talus at the base of the sea cliff (fig. 8.21). In most cases the structure can be small, especially if the fronting beach is composed of fine sand and has a low average slope. In

Figure 8.21 A low line of rip-rap has been successful in protecting the accumulated talus from wave attack during storms; the talus in turn shelters the sea cliff from rain wash and groundwater flow.

such circumstances the structure need only protect the talus from the run-up of weak waves that have lost nearly all of their energy in crossing the wide beach. By protecting the talus, the rip-rap reduces erosion on the cliff itself, because the talus shelters the cliff face from wind-driven rains and also reduces sapping from groundwater seepage. Growth of vegetation on the talus should be encouraged because it helps to stabilize the talus and thus protects the cliff from further erosion.

Rip current embayments that cut down the level of the beach can also play an important role in cliff erosion, particularly on steep coarse-sand beaches. In such circumstances, a more massive structure is required to withstand the stronger forces of waves that may break directly against it. To prevent undermining, the base of the structure must be placed below the lowest level the beach may reach when subjected to combined waves and rip currents; if possible, the base of the structure should rest on the bedrock beneath the beach sand.

Very high cliffs cannot be adequately protected by walls or other structures; no reasonably sized structure can protect the full elevation of the cliff. In that case, it is preferable to move a threatened dwelling back from the cliff edge. This approach usually costs less than building a seawall and provides the greatest certainty of success. The disposition of two houses threatened by erosion at Gleneden Beach will illustrate the point.

Following the erosion, the two dwellings projected well beyond the cliff edge (fig. 8.3). The one on the right was moved and is now safely beyond the reach of continued cliff recession. The dwelling on the left was not moved,

Figure 8.22 The homeowners' responses to the erosion at Gleneden Beach pictured in figure 8.3. One of two threatened houses was moved back from the cliff edge; the other was left in place and is now supported by I-beams, with railroad ties filling the space between them.

possibly because its concrete foundation was too massive. Instead, the portion of the house hanging over the cliff edge was supported by steel I-beams, and the spaces between were filled by railroad ties in order to reduce cliff erosion (fig. 8.22). Without a doubt, the result is the ugliest structure on the Northwest coast.

Most commonly, a structure, generally rip-rap, is placed at the base of a cliff to protect its toe and accumulated talus from wave attack. This leaves the upper portion of the cliff unprotected, and it continues to retreat due to rain wash, groundwater sapping, and human activities (graffiti carving, etc.). A number of schemes have been devised to protect bare upper cliff faces, including board walls and concrete gunite. These approaches have had mixed success in reducing the continued retreat of the bluff top.

Summary

Cliff erosion along the Northwest coast is extremely variable in extent and intensity. Particularly evident are spatial variations. For example, sea cliffs in the communities of Lincoln City and Gleneden Beach have undergone noticeable erosion during historic times, while cliffs in the Cannon Beach cell and at Bandon have experienced minimal wave erosion. On the other hand, the steep cliffs and offshore sea stacks at Cannon Beach and Bandon are clear evidence of extreme cliff retreat in the not too-distant past, perhaps initiated 300 years ago by the last subduction earthquake to affect the area. Subsequent uplift of the coast has reduced or eliminated continued erosion. We can expect renewed massive cliff erosion should there be another subduction earthquake.

A great deal of the spatial variability in sea cliff erosion is due to local factors that include the extent of the fronting beach, which determines its ability to act as a buffer between the waves and cliffs, and the composition of the bluffs, which governs their susceptibility to landsliding. Landslides are significant problems in some areas, the most notable example being the large Jump-Off Joe landslide in Newport, which is examined in detail in chapter 9. Cliff erosion has been episodic as well as spatially variable in areas such as Gleneden Beach, where the cliff is fronted by a coarse-sand beach.

Even in areas where sea cliffs are perceived to be undergoing significant erosion, however, the long-term retreat is small. In view of this, reasonable setback distances can be established that will keep new construction safe from continued cliff retreat. It is better to move threatened homes back from the cliff edge than to build protective structures, especially since the construction of seawalls and rip-rap revetments cannot guarantee adequate protection.

9 The Jump-Off Joe Fiasco

The rocky promontory called Jump-Off Joe was once one of the most picturesque spots on the Oregon coast (fig. 9.1). Legend has it that Joe, an Indian, jumped to his death while being pursued for a crime he had not committed. His lover, Mishi, who also jumped but survived, put a curse on the bluff. In view of subsequent events at Jump-Off Joe, the curse seems to have had its intended effect.

In 1942, a large landslide in the bluff at Jump-Off Joe carried more than a dozen homes to their destruction (Sayre and Komar 1988). In spite of continued slumping, a condominium was built on the remaining bluff in 1982. A certified geologist had determined that the site was stable even though it was adjacent to the 1942 landslide and in the area with the highest rate of erosion on the entire Oregon coast, and the Newport city government gave its approval to the project. Within three years, before the construction was even completed, slope retreat caused the foundation to fail, and the city ordered the destruction of the unfinished structure. The developers, the contractor, a lumber company, and the insurance company that had insured the project against slippage went bankrupt. Creditors with claims of $1 million were paid between 18 cents and 1 cent on the dollar. The consulting geologist lost his certification.

The debate over Jump-Off Joe was the most divisive land-use battle ever fought on the Oregon coast, and people still have strong feelings about the project. It was a classic confrontation between developers who thought their project would help a city grow and environmentalists who wanted to preserve the coastline. In the end, the issue was decided by Nature.

History of Erosion at Jump-Off Joe

Newport was founded in the 1860s by settlers who were attracted by the natural resources of the area, particularly the timber and abundant oysters.

Figure 9.1 The picturesque Jump-Off Joe sea arch inspired early tourists to pen lines of descriptive poetry. From the Oregon Historical Society, Portland.

The beauty of the coast also attracted tourists, who began to arrive in significant numbers in the 1890s.

One of the major tourist attractions in the Newport area was Jump-Off Joe, a rocky promontory just north of Nye Beach (figs. 9.1 and 9.2). Through the years Jump-Off Joe has been a much-photographed spot, and its rapid erosion is thus well documented (fig. 9.3). The earliest photographs, taken in the late 1800s, show the promontory still connected to the coast. Later photos show its separation and development into an arch. The arch eventually collapsed, and the resulting stacks continued to erode, so that today only small nubs remain, visible at low tide. After the loss of the original promontory, the name Jump-Off Joe was adopted for the area in general and has been used to refer to the landslide that developed in the 1940s as well as to the small remnant of terrace left behind as a promontory.

Development of the Jump-Off Joe area began in the early 1900s (Price 1975). Some landsliding endangered structures as early as 1921 (Baldwin 1985), but most of the damage occurred when a large slump developed over a period of months from late 1942 to spring 1943 (figs. 9.4 and 9.5). The slump is located between Sixth and Eleventh Streets, and the escarpment is west of and parallel to Coast Street (fig. 9.2). The 1942–43 slump involved about 15 acres and affected 15 houses. A few homes rode the slump block down intact and were occupied until 1966 (see figure 9.5). Eventually they were in danger of being undermined by wave erosion of the toe of the slide and were intentionally burned.

The *Yaquina Bay News* of March 11, 1943, made an interesting suggestion regarding the cause of the slump and earlier activity: "There was a formation of soapstone underneath and when the earth became saturated with water it would form a stream causing a crevice and pushing the ground up." The state geologist investigated the site a few weeks later and provided the earliest scientific account of the slump (Lowry and Allen 1945). The Jump-Off Joe bluff is a remnant of a marine terrace. Tertiary marine mudstones contained within the bluff are layered and dip steeply toward the sea (see fig. 8.10); most of the slumping takes place on shear zones within these mudstones.

The bluff retreat at Jump-Off Joe over the past century is documented in coastal charts and aerial photographs (Stembridge 1975c). Figure 9.2 shows the location of the cliff edge in 1868, 1939, and 1967. This diagram also shows that major slumping took place more than a century ago just north of the 1942–43 slump. The two slumps left a small segment of uneroded bluff between them, and it was this segment that became the site of condominium construction in 1982. Figure 9.2 indicates that the long-term sea cliff retreat

Figure 9.2 Cliff retreat at Nye Beach, Newport, from 1868 to 1976. Cliff edge lines were determined from old charts and aerial photographs (Stembridge 1975c). The black squares represent homes affected by the 1942–43 landslide.

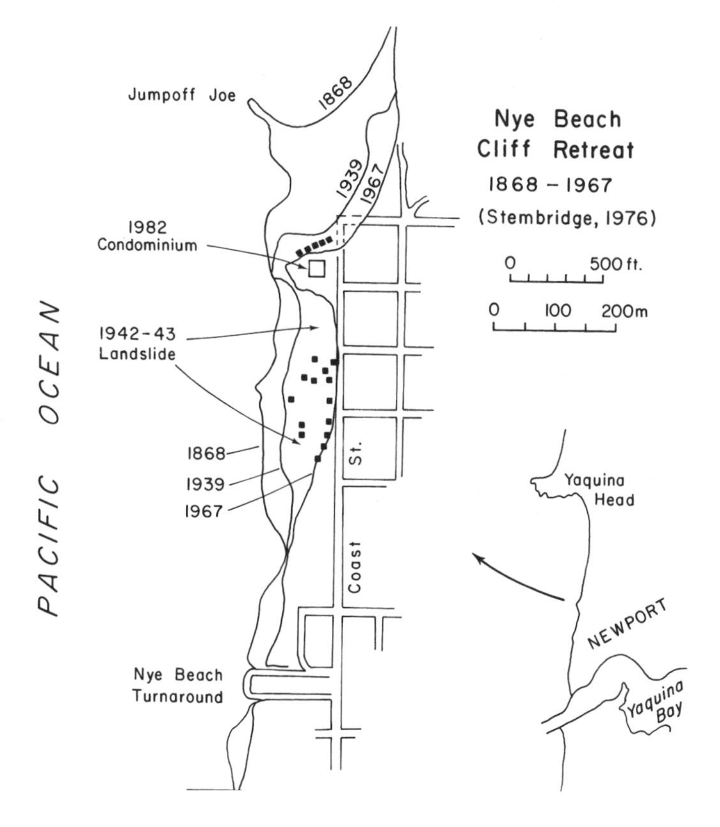

Figure 9.3 Photographs of Jump-Off Joe taken by tourists in 1880 (*top*), c. 1915 (*middle*), and 1978 (*bottom*). From the Lincoln County Historical Society, Newport.

was spatially variable but averaged several feet per year, a rate that is by far the highest on the Oregon coast.

An inventory of geological hazards along the Lincoln County coastline completed in 1975 gives an erosion rate of 7 feet per year for Jump-Off Joe and correctly concludes that such active landslides should remain undeveloped. This conclusion is ironic in view of the fact that the chief author of this report was to become the principal consulting geologist for the developers of Jump-Off Joe.

Figure 9.4 Photos taken on February 3, 1943, show some of the damage caused by the 1942–43 landslide at Jump-Off Joe. From the Lincoln County Historical Society, Newport, Oregon.

Figure 9.5 Aerial view of the 1942–43 landslide area in 1961. Some of the houses on the slump block were occupied until 1966. From the Lincoln County Historical Society, Newport, Oregon.

The Development of Jump-Off Joe

The story of condominium development at Jump-Off Joe begins in 1964 when the developers, Mr. and Mrs. Anderson of Newport, acquired the down-dropped block involved in the 1942–43 landslide and the adjacent uneroded bluff at the end of Eleventh Street (fig. 9.2; Sayre and Komar 1988). The city gave the Andersons these parcels in exchange for land to the north of the bluff.

The earliest geological investigation carried out for the developers, conducted by the well-known engineering firm of Shannon and Wilson, indicated that the down-dropped slump block was still active, as evidenced by fissures, its irregular hummocky topography, and back-tilted trees (see fig. 8.12). The investigators noted that wave erosion at the toe of the block was causing constant movement into the intertidal zone.

In spite of this reported slump activity and known high rates of erosion on the Jump-Off Joe bluff, the Andersons decided to go ahead with their plans for development. Grading and removal of vegetation on the down-dropped block began in December 1980 (fig. 9.6). Opposition to the project appeared along with the bulldozers. By mid-February 1981, the developers' attorney and geologist were meeting with neighboring homeowners to assure them of the appropriateness and benefits of the project.

Shannon and Wilson prepared a geotechnical report of the site for the developers that acknowledged the geological hazards at the site but proposed three measures to stabilize it:

1. A drain field to control groundwater seepage
2. Reduction of the steep slope paralleling Coast Street to a 1:2 slope using a combination of cut and fill
3. Construction of a seawall at the toe of the 1942–43 slump

On the basis of this report, a plan for the construction of 39 single-family homes was submitted to the Newport Planning Commission in early March 1981. Several opponents of the project also made presentations to the Planning Commission, arguing that the project endangered nearby private property, questioning the plans for reducing the landslide hazard, and stating that development should not be considered so close to the beach. In addition, a representative of the Oregon Land Conservation and Development Commission (LCDC) indicated that statewide land-use planning goals were not being satisfied. Oregon requires that all cities and counties have comprehensive land-use plans and that all plans conform to goals set by the LCDC (see chapter 10).

The Newport Planning Commission found the project attractive because it proposed new homes for a part of the city characterized by smaller, older homes, but postponed a decision on the subdivision. The next meeting of the Planning Commission was held in mid-April and focused on geological and geotechnical testimony from experts on both sides of the issue. The opponents were now represented by the Friends of Lincoln County (FLC), a group formed in the 1970s to oppose the development of wetlands in the Newport area. The FLC brought several letters from geologists and oceanog-

Figure 9.6 Grading on the 1942–43 Jump-Off Joe slump block in December 1980 in preparation for its development. From the Lincoln County Historical Society, Newport.

raphers that raised questions about the proposed hazards mitigation. An engineering geologist from Oregon State University questioned whether the developers' consultants had located the toe of the 1942–43 slump. If the failure zone was deeper than suspected and the toe was actually seaward of the proposed seawall, construction of the wall would further destabilize the slump block rather than providing support. Once again the Planning Commission postponed its decision.

The developers finally convinced the Planning Commission at a meeting in late April, and the project was given tentative approval as long as certain conditions were met. These included the completion of a detailed geotechnical study, an independent review of the developers' plans for stabilizing the block, and the establishment of beach access. In response, the FLC hired legal counsel and a professional geologist and asked the Newport City Council to review the Planning Commission's decision, alleging that the project violated state land-use goals. The FLC's attorney charged that the city government was unresponsive to the involvement of citizen groups in its decision-making procedures. A prodevelopment member of the Planning Commission and City Council characterized the FLC as combative and unwilling to compromise (Sayre and Komar 1988). The developers' attorney felt that too many conditions were placed on the developers at this stage of their plans and that the City Council took too long in ratifying the Planning Commission's decision. The City Council was trying to balance the opposing points of view and did not see any need for urgency. It did not complete its review until January 1982, more than six months later.

In the meantime, in May 1981, the Andersons advertised the property for sale. They were unable to find a buyer and continued with their development plans.

The detailed geotechnical study of the site requested of the developers by the Planning Commission was completed by Shannon and Wilson in July 1981 (Sayre and Komar 1988). Deeper drilling *did* reveal an older failure zone which had been active when the slump was much larger than present. At the city's request, the engineering firm CH_2M-Hill reviewed the report. Their resulting assessment noted that adding fill to reduce the slope along Coast Street would place a large load on the slump block, reducing its stability as well as occupying space originally planned for development. Cutting this slope would also require the purchase of private property and would expose a larger area to surface water erosion. They recommended that the developers take additional measures to stabilize the scarp.

The engineering report also placed the rate of erosion in the Jump-Off Joe area at several feet per year and expressed concern that the site and its seawall might become a peninsula over time, requiring the construction of wing walls. The designed seawall was not tall enough to stop overtopping by ocean waves, which would saturate the backfill and increase the weight the seawall was required to hold. Overtopping could also wash away some of the backfill. In addition, the report concluded that the footings were not deep enough to protect the structure from wave scour.

In early 1982, the developers applied to the Oregon Division of State Lands (DSL) for a permit to build a seawall, because Oregon's removal-fill law applied to the excavation and backfilling operations that would be involved in the construction. The DSL denied the application, stating that the seawall would produce only an illusion of safety in an area of known geological hazards, and that there would be no public benefit from its construction. The developers filed an appeal but later withdrew it.

In mid-January 1982, the City Council agreed with the Planning Commission's decision to allow the slump block to be resubdivided and developed, but then announced a few days later that it would reconsider its decision at a February 1 meeting. Most likely the council was going to postpone the decision once again because a cul-de-sac in the plan required a variance that had not been applied for. However, before the council could take that action, the Andersons suddenly withdrew their plans for development on the landslide itself and announced new plans to build 10 condominiums on the small remnant of bluff adjacent to the landslide. Their application for a building permit for that construction was granted a few days later.

A report written by the Andersons' consulting geologist was the first study prepared for the developers that focused on this small section of uneroded bluff. It was completed in the fall of 1981 while preliminary work was still under way on the down-dropped block. The report acknowledged the close proximity of massive landslides to the immediate north and south but concluded that the rate of cliff retreat was only 1 foot per year or less at the bluff itself, based on a comparison of aerial photographs taken in 1939 and 1972. The geologist did not explain the disagreement between this estimate and the 7-foot-per-year erosion rate given in the report he prepared for Lincoln County in 1975 (Rohleder et al. 1975), a rate that was confirmed by the 1981 CH_2M-Hill study. Based on his new lower rate of estimated erosion, the geologist established a setback line that would keep structures on the bluff safe from cliff retreat for 20 years. This setback line was followed in the later construction.

The 1981 report by the developers' geologist appears to have been critical in the City Council's decision to approve construction on the bluff (Sayre and Komar 1988). City planner Jan Monroe said that "If (the geologist) hadn't issued that report, they would never have given the project a building permit. If a person meets all the requirements and goes through the steps, they are issued a (building) permit. We have no discretionary authority to deny a permit based on gut feeling or knowing it's not good sense" (*Oregonian*, July 21, 1985, E10).

Shannon and Wilson reviewed the hazards report prepared by the geologist for the developers and suggested that a drainage system be installed. A 6-inch pipe was placed beneath the condominium to control groundwater saturation. It would later burst and accelerate erosion on the site.

The opponents of the development were unable to stop the construction of the condominium on the remnant of uneroded bluff. Building began in earnest in March 1982 (fig. 9.7), and by the end of the year all but the interior was completed. The precarious position of the building, on a rapidly eroding bluff with landslides on both sides (fig. 9.8), should have been ample warning to potential buyers. Opponents of the development who lived near the construction site placed signs in their front yards as an additional warning of the landslide hazard (fig. 9.9).

Each unit was to sell for $250,000, but sales were slow. The early 1980s was a time of high interest rates and a depressed real estate market. Construction was halted in December 1982 before the interior was completed. The developers had been unable to obtain a construction loan and ran out of money. Most of the subcontractors had placed liens on the condominium. An appraisal placed the value of the unfinished project at $1 million (Sayre and Komar 1988).

As early as September 1981, the Andersons had stopped making payments on a loan for the subdivision project, although this was not known publicly until near the end of 1982. Accumulated interest during the delay and demands by their lending institution ultimately led in late 1982 to foreclosure and auction of the down-dropped landslide block. The land was sold to the

Figure 9.7 Condominiums under construction in 1981 on the terrace remnant at Jump-Off Joe. From the *Newport News-Times.*

Figure 9.8 This site on a rapidly eroding bluff with landslides on both sides was extremely precarious for development. The southwestern portion of the structure (lower right) is already beginning to tilt.

bank for more than $850,000. Within a year, this bank found itself in trouble because of poor loan practices and was forced to merge with another bank.

The Andersons filed for bankruptcy in May 1983. Just prior to that, they purchased insurance against slippage of the condominiums. The insurance premiums were paid by a committee of the 19 secured creditors who held liens on the construction; among this group was the city of Newport (Sayre and Komar 1988).

By September 1984, sloughing of the bluff had undermined the perimeter fence around the condominium. The drainage pipe burst, probably due to

Figure 9.9 The Friends of Lincoln County erected lawn signs in the Jump-Off Joe area to discourage buyers. From the *Newport News-Times.*

Figure 9.10 The foundation of this condominium failed during renewed slumping in 1985. The stress placed on the building by the ground movement and loss of support caused the windows to shatter.

Figure 9.11 The final demolition of the condominiums in October 1985 brought to an end the contention over developing the Jump-Off Joe landslide site. From the *Newport News-Times.*

slippage, exacerbating the problem. A larger slump developed on the remnant bluff, causing the foundation to fail (fig. 9.10). Slump movement was not directly seaward, but had a southerly component, suggesting that regrading of the 1942–43 slump surface during development may have been a contributing factor. In January 1985, the city ordered the demolition of the condominiums, and they were torn down later that year (fig. 9.11). The salvager paid the city $4,000.

The developers filed a $375,000 claim with the insurance company, but the claim was not settled for more than a year because the insurance company had also filed for bankruptcy. By the time the company was ready to investigate the claim, the condominium had been destroyed by the city. In the end, the insurance company paid out $225,000. After administrative expenses, legal fees, and other costs were subtracted, there was only $131,000 left to meet the secured creditors' claims, which totaled $720,000 (Sayre and Komar 1988). The largest settlements went to the contractor and the lumber company, both of which were also bankrupt. The 43 unsecured creditors, including the developers' attorney and their consulting geologist, requested a total of $283,000 but received only $3,544.

The Oregon State Board of Geologist Examiners filed a complaint against the developers' geologist over this and five other projects (Sayre and Komar 1989). The board decided to revoke his certification, citing in a news release his "incompetence and gross negligence." The Newport City Council adopted a new subdivision ordinance and a new comprehensive plan. Friends of Lincoln County was involved in the proceedings and contended that no development should be allowed at Jump-Off Joe. Nevertheless, the area remains zoned for high-density multifamily dwellings, although now with a geological hazards overlay that allows the city to request additional information and more exploration. The ownership of the bluff is still in question, but the down-dropped block is owned by a Los Angeles developer. The city hopes eventually to acquire the land.

10 The Northwest Coast—A Heritage to Be Preserved

As the twenty-first century approaches, we face new challenges on the Northwest coast. Behind us is a century of development, and now is a good time to reflect on where we are and what must be done to manage and preserve our coastal resources. We have a much better understanding of the processes of coastal erosion and its impacts than we did even a decade ago. Will it be possible to use that information to make management decisions that will limit erosion and financial loss on the coast? Can midcourse corrections help us avoid further proliferation of the seawalls and rip-rap revetments that have begun to mar our coast? The decision is ours to make. Will it be seawalls or wise management?

We are fortunate in having established public ownership and good access to the beaches of the Northwest. On the Washington coast, the 25-mile stretch of shore on the Long Beach Peninsula is open to the public along its full length. Farther to the north is Olympic National Park, with complete public access to beaches and trails along the rocky stretches of coast. Even in areas of the Washington coast that are privately owned, the public has access to the beaches. This is also the case along the Oregon coast, where an unparalleled system of state parks, waysides, and paths provides more than 600 beach access points.

We in the Northwest tend to take for granted our ready access to any beach. We come to more fully appreciate that freedom only when we travel to other states where those rights have not been established. For example, only 10 miles of the 1,300-mile coast of Massachusetts are publicly owned (Straton 1977). Most of the Florida coast has been claimed by hotels, exclusive beach clubs, and private homeowners. Less than one-fifth of the 1,200 miles of coast in California is open to the public.

The public ownership of beaches in the Northwest came about in part through happenstance. The first white settlers lived in relative isolation along the shores of bays and estuaries (see chapter 4). North-south travel

Figure 10.1 The beaches were the best routes for north-south travel along the Northwest coast prior to the construction of highways. From the Oregon Historical Society, Portland.

was easiest along the wide ocean beaches, which were adopted as ready-made roads (fig. 10.1). This use was formalized in 1899 when the Oregon legislature designated the 30 miles of ocean beach between the Columbia River and the south line of Clatsop County a public highway (Straton 1977). In 1913, the law was amended to make all ocean beaches highways. This placed beach ownership firmly in public hands, or so people thought until the right of full access to the dry-sand beach was challenged during the 1960s.

Figure 10.2 Recreation on Northwest beaches over the decades was important in establishing legal public access. From the Oregon Historical Society, Portland.

Figure 10.3 Former governor Tom McCall surveying the log barricade placed on the sandy beach in front of the Surfsand Motel in Cannon Beach on May 13, 1967. This denial of public access to the beach eventually resulted in the passage of the Oregon Beach Bill. From the *Oregonian,* May 1967.

The designation of beaches as highways was limited by law to the area extending from extremely low tide landward to "ordinary" high tide. Most people assumed that this meant the entire sand beach and became accustomed to unquestioned use of it as a recreational playground (fig. 10.2). Accordingly, there was a great deal of public concern during the summer of 1966 when the owner of the Surfsand Motel in Cannon Beach built a low barricade of logs around the dry-sand area adjacent to the motel and erected signs stating "Surfsand Guests Only Please" (fig. 10.3). This challenge to public ownership and access resulted in the passage of the Beach Bill in 1967, which established the public's right by long-established custom to recreational use of the dry sands of beaches along Oregon's coast.

Immediately after the Beach Bill became law, the state Highway Commission began a survey of the entire coast to establish a permanent beach zone line, and the coordinates of that survey line became an integral part of the law. The line, which roughly corresponds to the edge of the vegetation backing the sandy beach, remains today as the jurisdictional limit of the state's easement with rights of its citizens for recreational use.

The first legal challenge to the Beach Bill came not from the Surfsand Motel in Cannon Beach, but rather involved an attempt to build a private road across the beach at Neskowin (fig. 10.4). Built to provide access to property along the bluff to the south on Cascade Head, the road extended in a U across the sandy beach, approximately 200 feet beyond the newly established beach zone line. The state challenged the road in the courts. The judgment, handed down on August 26, 1968, ruled that the road was not in the public interest and affirmed that the public had acquired an easement. The Surfsand Motel case was similarly decided in 1969.

State jurisdiction over the beaches further requires that coastal property owners obtain permits before constructing seawalls or rip-rap revetments that would extend across the beach zone line. Denial of such permits has been one of the main factors limiting the proliferation of shore protection structures along the Oregon coast.

If the public has the right to use the beaches, it must also be guaranteed access to them. Fortunately, state officials recognized this at an early date, and in 1964 the Oregon State Parks Division began a program to improve and guarantee access by acquiring parcels of land at approximately 3-mile intervals. These now constitute our numerous state parks, waysides, and pathways leading to the beach.

Although our rights to reach and use beaches have been preserved, our shores are not yet safe. Rapacious land developers and urban sprawl have become a serious threat to the Northwest coast. Many a once-quiet seashore resort has been lost to excessive development. Urban sprawl is insidious along the coast. The growth tends to follow Highway 101, which is pressed between the sea and the mountains of the Coast Range. Each year, more of the green space between successive communities vanishes as the highway is lined with ever more RV parks, motels, and fast-food restaurants. During the height of the tourist season, parts of Highway 101 feel like downtown Los Angeles. Nowhere is this worse than in Lincoln City, which has grown into a hodgepodge of unplanned and unattractive commercial develop-

Figure 10.4 Construction of a road across the beach at Neskowin triggered the first court test of Oregon's Beach Bill. From the Oregon Department of Transportation.

ments. Touted by the Chamber of Commerce as the "Twenty Miracle Miles," visitors more often glumly refer to it as the "twenty miserable miles."

The urban sprawl extends all the way to the beaches. The sea cliffs and sand spits are lined with houses, motels, and condominiums, interrupted only by state parks. The wholesale development of properties backing the beaches has resulted in increased problems with erosion because natural processes continue to act without regard to humans' presence. The erosion has led to the proliferation of seawalls and revetments. The natural beauty of the coast is increasingly marred by the presence of these structures. This is where the challenge now lies if we are to preserve the beauty and naturalness of the Northwest coast for future generations.

Increasing development and its associated problems have led nearly all the coastal states to adopt plans for the management of their coastal zones. While these programs vary in their emphasis, most reflect a concern for the preservation of coastal environments. The main initiative has been the Coastal Zone Management Act passed by the U.S. Congress in 1972 with four main objectives: (1) to protect fragile coasts; (2) to minimize life and property losses from coastal hazards; (3) to create better conditions for coastal resource use, including better access for recreation; and (4) to promote intergovernmental cooperation, leading to a reduction in bureaucracy. The federal government provided grants to the states to establish coastal management programs, and the states were required to submit their programs to the federal government for evaluation and approval.

In 1976, Washington became the first state to have its coastal management program approved and implemented; approval of Oregon's program followed soon thereafter in 1977. Both states delegated much of the development of their coastal zone management programs to the counties and cities. This has turned out to be an unfortunate decision. Each community has somewhat different regulations, and too often the county or city officials are lax in enforcing the management policies. A tragic example is the attempt to develop the Jump-Off Joe landslide site discussed in chapter 9.

One of the basic principles of coastal zone management is to avoid placing homes and other structures in hazardous zones. Jump-Off Joe was by far the best-known geological hazard along the Oregon coast, yet approval to develop was forthcoming at the local level, from the city of Newport. The consequences of this unwise development became obvious when the still-unfinished condominium was destroyed by slumping and bluff erosion.

An enlightened and well-implemented coastal zone management program should prevent construction in areas prone to rapid erosion and landslides. We don't need any more Jump-Off Joe fiascoes. More uncertain is the regulation of development close to cliff edges that are retreating at slow rates from episodic erosion. In most cases it is difficult to predict the rate and degree of future erosion, and the wisest policy is to stay well back from the cliff edge. The establishment of reasonable setback lines based on long-term recession rates of the sea cliff would do much to ensure building safety.

Setback lines are particularly important on sand spits and in other dune areas because of the rapidity with which erosion processes cut back such areas. Our analyses of predevelopment aerial photographs of Siletz Spit (see chapter 6) showed that erosion could cut back the foredunes by up to 50 feet within just a few weeks. Had setback lines been established on Siletz Spit, development in that naturally ephemeral zone could have been prevented. Unfortunately, the developers did not commission studies of the spit's stability before building, and homes were placed in foredune areas that had eroded away only a decade earlier.

Potential increases in sea level and subsidence associated with subduction earthquakes should also be considered when drawing setback lines, both for sea cliffs and for foredunes backing beaches. New homes could be designed to be more readily movable should erosion be greater than expected.

Setback lines have been established for many areas of the Oregon coast, yet they are commonly ignored and development in hazardous areas has not been significantly curtailed. The Siletz littoral cell is one of the most erosion-prone areas on the Oregon coast, and one of the most extensively developed (Good 1994). Nearly 20 percent of the construction there has oc-

Figure 10.5 Annual construction of shore protection structures in the 14-mile-long Siletz littoral cell and their cumulative length (6.8 miles as of 1991). Construction increases immediately after El Niño events, which can be ranked as moderate (M), strong (S), or very strong (VS) depending on the intensity of the coastal erosion processes they generate. From Good 1994.

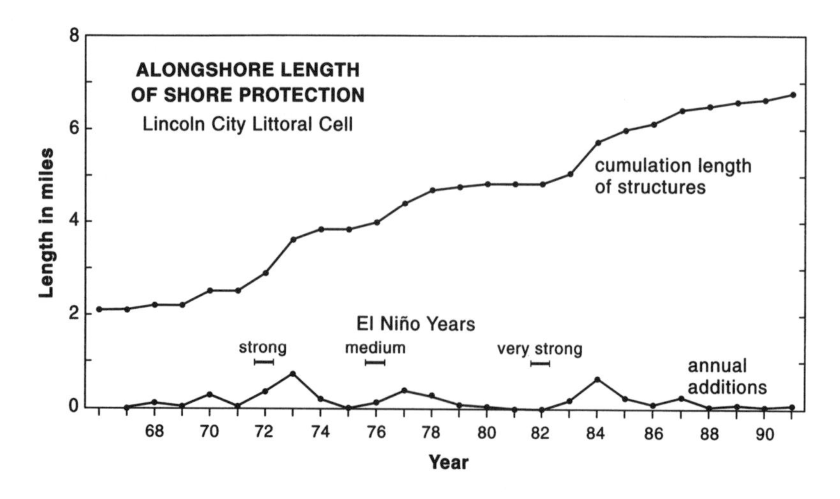

Figure 10.6 A massive rip-rap revetment on the beach in the Siletz littoral cell restricts access to the beach and is a visual blight.

curred since Oregon's coastal zone management program was adopted and went into effect in 1977. The spit's shoreline has been gradually hardened with protective structures, mostly large rip-rap revetments and concrete seawalls (fig. 10.5; Good 1994).

By 1991, 6.8 miles, or 49 percent, of the 14 miles of beachfront shoreline within the Siletz littoral cell had some sort of protective structure. A review of the permits that were issued for these structures showed that in 35 percent of the cases there was no threat from erosion or land instability that actually warranted building a shore protection structure (Good 1994). In 28 percent of the cases the lots were vacant, so no homes could have been in danger.

Local policies require property owners to install a hard shore protection structure in order to obtain a building permit. A number of vacant oceanfront lots are shallow and virtually unbuildable without an erosion prevention structure, but new construction can ignore the setback zone so long as the property can be adequately "stabilized." Too often, this means that a new seawall or revetment is constructed whether or not there is any danger from active or potential erosion. The end result is that much of the beauty of the coast within the Siletz littoral cell, and especially within the area of Lincoln City, is marred by large seawalls and great heaps of rip-rap (fig. 10.6).

In addition to their aesthetic effects, shore protection structures block new sand from reaching the beach. As more and more structures are constructed, the supply of sand will further diminish, and "protected" beaches will become ever narrower as the sea level continues to rise. Coastal management policies were intended to give preference to hazard avoidance and nonstructural means of erosion control. In practice, however, as seen in the Siletz littoral cell, seawalls and revetments have become the preferred hazard-reduction strategy, much to the detriment of the coast (Komar and Good 1989b; Good 1994).

Figure 10.7 This massive creosote-covered seawall is an unnecessary eyesore; there was minimal bluff erosion at this site.

Too often, those who decide whether or not a property should be developed or whether the construction of a shore protection structure is warranted do not adequately assess the property's susceptibility to erosion. Such an assessment requires determining the expected wave run-up during extreme storms at high tides versus the capacity of the fronting beach to act as a buffer (see chapter 3). This information must be determined for each oceanfront lot that is to be developed or protected. Detailed site evaluations by registered engineers or geologists are required when a development is proposed, but their reports almost never include analyses of the potential susceptibility to erosion.

It is surprising how many expensive seawalls and revetments are built to protect areas that do not need them. Such is true of a massive seawall on the scenic north Oregon coast (fig. 10.7) that creates both a visual blight and an olfactory one—it can be smelled for a considerable distance because it is coated with creosote. In this case, the developer was told to build the home well back from the cliff edge so there would be no danger from erosion. This would have involved a relatively small setback distance since the long-term erosional retreat of the cliff at that site has been negligible. Unfortunately, this reasonable recommendation was ignored, and the house was placed as close to the cliff edge as possible. As soon as the house was completed the owner requested permission to build a seawall, and it was granted even though he had ignored the previous warning and no erosion had occurred subsequent to construction.

In another example, permission was given to construct a large seawall in front of an older house (fig. 10.8), in spite of objections from neighbors and even though the bluff has not experienced erosion within historic times. The absence of erosion was readily apparent in the thick vegetation covering the bluff, including sizable trees that could also be seen on old photographs of the site. The geologist who performed the site inspection for the homeowner reported that the property was undergoing 2 feet of bluff retreat per year. This would seem to be another case of a consultant tailoring his conclusions to fit the desires of his client.

Unfortunately, these two cases are not unique. County commissions and city councils almost always grant requests for the construction of seawalls, whether or not they are needed to halt erosion. Only the state has attempted to prevent seawall construction, but the state has a say in the matter only if the proposed structure crosses the beach zone line or if a large volume of material is to be excavated during construction.

A new element that must be considered in management decisions on the Northwest coast has been recognized within the last decade: the potential for a subduction earthquake and accompanying tsunami. Only recently have we learned that such events have already occurred a number of times in the past, the last one on the evening of January 26, 1700 (see chapter 2). Based on the time intervals between past events, the 300 years since the last major earthquake places us well within the window for another. We do not know whether it will occur today, tomorrow, or not for another century or more, but without a doubt such an event is on the agenda for the Northwest coast, and we must be prepared for it.

Figure 10.8 An unnecessary seawall now mars the natural beauty of the coast south of Cannon Beach, Oregon.

The next subduction earthquake will probably have a magnitude on the order of 8 to 9 and will cause mass destruction along the coast and well inland, effectively isolating communities from outside help. It is estimated that the tsunami waves generated offshore by the earthquake will reach the shore within 10–20 minutes. According to studies of sediments deposited by past tsunamis (chapter 2), these immense waves may wash over low-lying coastal areas and travel many miles up bays and estuaries, causing destruction all along their shores. It is imperative that residents disregard the earthquake shaking and destruction and get out of the path of the expected tsunami. This danger is being recognized by more and more coastal communities, and a few already have evacuation plans, including warning sirens and evacuation routes that lead to safe elevated areas. A number of schools teach their students about earthquake and tsunami hazards, and a few have annual drills.

Unfortunately, the extreme tsunami hazard has not yet significantly affected development along the coast. It is expected that the primary response will be to place schools and hospitals in safer areas and in buildings with reinforced construction. Little consideration has been given to the thousands of homes—and people—that will be in the path of the destructive tsunami waves.

It would be easy to become despondent in the face of the rapidly expanding development and the proliferation of shore protection structures on the scenic Northwest coast. It would be easy to give up and let developers and other property owners do exactly as they wish. Yet a midcourse correction that could lead to the preservation of the remaining natural areas of the coast is still possible. Unwise development can be prevented, but it will require basic changes in our attitudes toward development, changes in shoreline management policies to incorporate our present understanding of erosion processes, the courage of government officials to make controversial decisions, and, most of all, the support of the public who wants to preserve the heritage of the Northwest coast.

References

Adams, J. 1984. Active deformation of the Pacific Northwest continental margin. *Tectonics* 3(4): 449–72.

Aguilar-Tunon, N. A., and P. D. Komar. 1978. The annual cycle of profile changes of two Oregon beaches. *Ore Bin* 40: 25–39. The *Ore Bin* is published by the state Department of Geology and Mineral Industries, Portland, Ore.

Allen, J. E., M. Burns, and S. C. Sargent. 1986. *Cataclysms on the Columbia.* Portland, Ore.: Timber Press.

Allyn, S. 1975. Bayocean, a washed out dream. In *Northwest,* the *Sunday Oregonian* magazine, April 6, 1975, 6.

Atwater, B. F. 1987. Evidence for great Holocene earthquakes along the outer coast of Washington State. *Science* 236: 942–44.

Atwater, B. F., and D. K. Yamaguchi. 1991. Sudden, probably coseismic submergence of Holocene trees and grass in coastal Washington State. *Geology* 19: 706–9.

Atwater, T. 1970. Implications of plate tectonics for the Cenozoic tectonic evolution of western North America. *Bulletin of the Geological Society of America* 81: 3513–36.

Baker, V. R. 1973. *Paleohydrology and Sedimentology of Lake Missoula Flooding in Eastern Washington.* Geological Society of America Special Paper 144.

Baldwin, E. M. 1945. Some revisions of the Late Cenozoic stratigraphy of the southern Oregon coast. *Journal of Geology* 52: 35–46.

———. 1985. *Geology of Oregon.* Dubuque, Iowa: Kendall-Hunt.

Ballard, R. L. 1964. Distribution of beach sediment near the Columbia River. Master's thesis, University of Washington; University of Washington, Department of Oceanography, Tech. Report No. 98.

Barnett, T. P. 1984. The estimation of "global" sea level change: a problem of uniqueness. *Journal of Geophysical Research* 89(C5): 7980–88.

Bascom, W. 1980. *Waves and Beaches.* 2d ed. Garden City, N.Y.: Anchor Books.

Beckham, S. D. 1977. *The Indians of Western Oregon.* Coos Bay, Ore.: Arago Books.

Boggs, S. 1969. Distribution of heavy minerals in the Sixes River, Curry County, Oregon. *Ore Bin* 31: 133–50.

Boggs, W. S., and C. A. Jones. 1976. Seasonal reversal of flood-tide dominated sediment transport in a small Oregon estuary. *Bulletin of the Geological Society of America* 87: 419–26.

Borden, C. E. 1979. Peopling and early cultures of the Pacific Northwest. *Science* 203: 963–71.

Boxberger, D. L., and D. McFeron. 1983. Building Highway 101. *Oregon Coast* (October–November 1983): 30–31.
Bretz, J. H. 1969. The Lake Missoula floods and the Channeled Scabland. *Journal of Geology* 77: 505–43.
Brooks, H. C. 1979. Plate tectonics and the geologic history of the Blue Mountains. *Oregon Geology* 41: 71–80.
Brown, H. E., G. D. Clark, and R. J. Pope. 1958. Closure of the breach in Bayocean Peninsula, Oregon. *Journal of Waterways and Harbors Division* 84: 1–20.
Byrne, J. V. 1963. Coastal erosion, northern Oregon. In *Essays in Marine Geology in Honor of K. O. Emery,* ed. T. Clements, 11–33. Los Angeles: University of Southern California Press.
———. 1964. An erosional classification for the northern Oregon coast. *Annals of the Association of American Geographers* 54: 329–35.
Byrne, J. V., G. A. Fowler, and N. J. Maloney. 1966. Uplift of the continental margin and possible continental accretion off Oregon. *Science* 154: 1654–56.
CH_2M-Hill. 1981. Review of geotechnical report for Beachwood Estates project at Jump-Off Joe, Newport, Oregon. Unpublished technical report.
Chan, M. A., and R. H. Dott Jr. 1983. Shelf and deep-sea sedimentation in Eocene forearc basin, western Oregon: Fan or non-fan? *Bulletin of the American Association of Petroleum Geologists* 67: 2100–16.
Clemens, K. E., and P. D. Komar. 1988a. Oregon beach-sand compositions produced by the mixing of sediments under a transgressing sea. *Journal of Sedimentary Petrology* 58: 519–29.
———. 1988b. Tracers of sand movement on the Oregon coast. In *Proceedings of the 21st Coastal Engineering Conference,* American Society of Civil Engineers, 1338–51.
Cooper, W. S. 1958. *Coast Dunes of Oregon and Washington.* Geological Society of America Memoir 72.
Creech, C. 1981. Nearshore wave climatology, Yaquina Bay, Oregon (1971–1981). Oregon State University Sea Grant Program, Report ORESU-T-81-002.
Curray, J. R. 1965. Late Quaternary history, continental shelfs of the United States. In *The Quaternary of the United States,* ed. H. E. Wright and D. G. Frey, 723–35. Princeton: Princeton University Press.
Darienzo, M. E., and C. D. Peterson. 1990. Episodic tectonic subsidence of late Holocene salt marshes, northern Oregon central Cascadia margin. *Tectonics* 9: 1–22.
———. 1995. Magnitude and frequency of subduction-zone earthquakes along the northern Oregon coast in the past 3,000 years. *Oregon Geology* 57(1): 3–12.
Darienzo, M. E., C. D. Peterson, and C. Clough. 1994. Stratigraphic evidence for great subduction-zone earthquakes for four estuaries in northern Oregon, U.S.A. *Journal of Coastal Research* 10: 850–76.
Defant, A. 1958. *Ebb and Flow.* Ann Arbor: University of Michigan Press.
Dicken, S. N. 1961. Some recent physical changes of the Oregon coast. University of Oregon, Report to the Office of Naval Research, November 15, 1961.
———. 1971. *Pioneer Trails of the Oregon Coast.* Portland: Oregon Historical Society.
Dodds, G. B. 1986. *The American Northwest: A History of Oregon and Washington.* Arlington Heights, Ill.: Forum Press.
Douthit, N. 1986. *Oregon South Coast History.* Coos Bay, Ore.: River West Books.
Drake, E. T. 1982. Tectonic evolution of the Oregon continental margin. *Oregon Geology* 44(2): 15–21.
Drucker, P. 1963. *Indians of the Northwest Coast.* Garden City, N.Y.: Natural History Press.
Enfield, D. B., and J. S. Allen. 1980. On the structure and dynamics of monthly mean sea

level anomalies along the Pacific coast of North and South America. *Journal of Physical Oceanography* 10: 557–78.

Espy, W. R. 1977. *Oysterville.* New York: Clarkson N. Potter.

Flick, R. E., and D. R. Cayan. 1985. Extreme sea levels on the coast of California. *Proceedings of the 18th Coastal Engineering Conference,* American Society of Civil Engineers, 886–98.

Fox, W. T., and R. A. Davis. 1978. Seasonal variation in beach erosion and sedimentation on the Oregon coast. *Bulletin of the Geological Society of America* 89: 1541–49.

Gibbs, Jim. 1978. *Oregon's Salty Coast.* Seattle: Superior.

Goldfinger, C., L. D. Kulm, R. S. Yeats, B. Appelgate, M. E. MacKay, and G. F. Moore. 1992. Transverse structural trends along the Oregon convergent margin: Implications for Cascadia earthquake potential and crustal rotations. *Geology* 20: 141–44.

Good, J. W. 1994. Shore protection policy and practices in Oregon: An evaluation of implementation success. *Coastal Management* 22: 325–52.

Gornitz, V., L. Lebedeff, and J. Hansen. 1982. Global sea level trend in the past century. *Science* 215: 1611–14.

Guza, R. T., and E. B. Thornton. 1981. Wave set-up on a natural beach. *Journal of Geophysical Research* 86(C5): 4133–37.

———. 1982. Swash oscillations on a natural beach. *Journal of Geophysical Research* 87(V1): 483–91.

Hall, R. L., and S. Radosevich. 1995. Episodic flooding of prehistoric settlements at the mouth of the Coquille River. *Oregon Geology* 57(1): 18–22.

Hamilton, S. F. 1973. *Oregon Estuaries.* Salem: Oregon Division of State Lands.

Hanneson, B. 1962. Changes in the vegetation on coastal dunes in Oregon. Master's thesis, University of Oregon.

Heller, P. L., and W. R. Dickinson. 1985. Submarine ramp facies model for delta-fed, sand-rich turbidite systems. *Bulletin of the American Association of Petroleum Geologists* 69: 960–76.

Hicks, S. D. 1978. An average geopotential sea level series for the U.S. *Journal of Geophysical Research* 83(C3): 1377–79.

Hicks, S. D., H. A. Debaugh, and L. E. Hickman. 1983. *Sea Level Variations for the United States, 1855–1980.* Rockville, Md.: U.S. Department of Commerce, National Oceanic and Atmospheric Administration, National Ocean Service.

Hoffman, J. S., D. Keyes, and J. G. Titus. 1983. *Projecting Future Sea Level Rise: Methodology, Estimates to the Year 2100, and Research Needs.* Washington, D.C.: Environmental Protection Agency.

Holman, R. A., and A. H. Sallenger. 1985. Set-up and swash on natural beach. *Journal of Geophysical Research* 90(C1): 945–53.

Huyer, A., W. E. Gilbert, and H. L. Pittock. 1983. Anomalous sea levels at Newport, Oregon, during the 1982–83 El Niño. *Coastal Oceanography and Climatology News* 5: 37–39.

Kelsey, H. M. 1990. Late Quaternary deformation of marine terraces on the Cascadia subduction zone near Cape Blanco, Oregon. *Tectonics* 9: 983–1014.

Komar, P. D. 1997. *Beach Processes and Sedimentation.* 2nd edition. Upper Saddle River, N.J.: Prentice-Hall.

———. 1978a. Beach profiles on the Oregon and Washington coasts obtained with an amphibious DUKW. *Shore and Beach* 46: 26–33.

———. 1978b. Wave conditions on the Oregon coast during the winter of 1977–78 and the resulting erosion of Nestucca Spit. *Shore and Beach* 44: 3–8.

———. 1983. The erosion of Siletz Spit, Oregon. In *Coastal Processes and Erosion,* 65–76. Boca Raton, Fla.: CRC Press.

———. 1986. The 1982–83 El Niño and erosion on the coast of Oregon. *Shore and Beach* 54: 3–12.

Komar, P. D., and J. W. Good. 1989a. Long-term erosion impacts of the 1982–83 El Niño on the Oregon coast. In *Coastal Zone '89*, 3785–94, American Society of Civil Engineers.

———. 1989b. The Oregon coast in the twenty-first century: A need for wise management. In *Ocean Agenda 21*, ed. C. L. Smith. Oregon State Sea Grant, ORESU-B-89-001.

Komar, P. D., J. W. Good, and S.-M. Shih. 1988. Erosion of Netarts Spit, Oregon: Continued impacts of the 1982–83 El Niño. *Shore and Beach* 57: 11–19.

Komar, P. D., and R. A. Holman. 1986. Coastal processes and the development of shoreline erosion. *Annual Reviews of Earth and Planetary Sciences* 14: 237–65.

Komar, P. D., and M. Z. Li. 1991. Black-sand placers at the mouth of the Columbia River, Oregon and Washington. *Marine Mining* 10: 171–87.

Komar, P. D., J. R. Lizarraga, and T. A. Terich. 1976a. Oregon coast shoreline changes due to jetties. *Journal of Waterways, Harbors and Coastal Engineering Division*, 102: 13–30.

Komar, P. D., and B. A. McKinney. 1977. The spring 1976 erosion of Siletz Spit, Oregon, with an analysis of the causative storm conditions. *Shore and Beach* 45: 23–30.

Komar, P. D., W. Quinn, C. Creech, C. C. Rea, and J. R. Lizarraga-Arcineiga. 1976b. Wave conditions and beach erosion on the Oregon coast. *Ore Bin* 38: 103–12.

Komar, P. D., and C. C. Rea. 1976. Erosion of Siletz Spit, Oregon. *Shore and Beach* 44: 9–15.

Komar, P. D., and S.-M. Shih. 1991. Sea-cliff erosion along the Oregon coast. In *Coastal Sediments '91*, 1558–70. American Society of Civil Engineers.

———. 1993. Cliff erosion along the Oregon coast: A tectonic–sea level imprint plus local controls by beach processes. *Journal of Coastal Research* 9: 747–65.

Komar, P. D., and T. A. Terich. 1976. Changes due to jetties at Tillamook Bay, Oregon. In *Proceedings of the 15th Coastal Engineering Conference*, 1791–1811. American Society of Civil Engineers.

Komar, P. D., R. W. Torstenson, and S.-M. Shih. 1991. Bandon, Oregon: Coastal development and the potential for extreme ocean hazards. *Shore and Beach* 59: 14–22.

Kulm, L. D., and J. V. Byrne. 1966. Sedimentary response to hydrography in an Oregon estuary. *Marine Geology* 4: 85–118.

Kulm, L. D., and G. A. Fowler. 1974. Oregon continental margin structure and stratigraphy: A test of the imbricate thrust model. In *The Geology of Continental Margins*, ed. C. A. Burk and C. L. Drake, 261–83. New York: Springer-Verlag.

Lowry, W. D., and J. E. Allen. 1945. An investigation of the sea-cliff subsidence of March 30, 1943, at Newport, Oregon. *Geological Society of the Oregon Country Newsletter* 2(15): 99.

McCaffrey, R., and C. Goldfinger. 1995. Forearc deformation and great subduction earthquakes: Implications for Cascadia offshore earthquake potential. *Science* 267: 856–59.

McKee, B. 1972. *Cascadia: The Geological Evolution of the Pacific Northwest.* New York: McGraw-Hill.

McKinney, B. A. 1977. The spring 1976 erosion of Siletz Spit, Oregon, with an analysis of the causative wave and tide conditions. Master's thesis, Oregon State University.

Meier, Gary. 1991. When the highway was sand. *Oregon Coast* (January–February 1991): 26–29.

Mitchell, C. E., P. Vincent, R. J. Weldon, and M. A. Richards. 1994. Present-day vertical deformation of the Cascadia margin, Pacific Northwest, United States. *Journal of Geophysical Research* 99(B6): 12257–77.

Muhs, D. R., H. M. Kelsey, G. H. Miller, G. L. Kennedy, J. F. Whelan, and G. W. McInelly.

1990. Age estimates and uplift rates for Late Pleistocene marine terraces: Southern Oregon portion of the Cascadia forearc. *Journal of Geophysical Research* 95(B5): 6685–98.
National Research Council. 1985. *Glaciers, Ice Sheets, and Sea Level.* Washington, D.C.: Polar Research Board.
Nokes, J. R. 1991. *Columbia's River: The Voyages of Robert Gray, 1787–1793.* Tacoma: Washington State Historical Society.
North, W. B. 1964. Coastal landslides in northern Oregon. Master's thesis, Oregon State University.
North, W. B., and J. V. Byrne. 1965. Coastal landslides in northern Oregon. *Ore Bin* 27(11): 217–41.
O'Donnell, T. 1988. *That Balance So Rare: The Story of Oregon.* Portland: Oregon Historical Society.
Orr, E. L., W. N. Orr, and E. M. Baldwin. 1992. *Geology of Oregon.* Dubuque, Iowa: Kendall-Hunt.
Peterson, C. D., and G. R. Priest. 1995. Preliminary reconnaissance survey of Cascadia paleotsunami deposits in Yaquina Bay, Oregon. *Oregon Geology* 57(2): 33–40.
Peterson, C., K. F. Scheidegger, and P. D. Komar. 1982. Sand-dispersal patterns in an active-margin estuary of the northwestern United States as indicated by sand composition, texture and bedforms. *Marine Geology* 50: 77–96.
Peterson, C., K. Scheidegger, P. D. Komar, and W. Niem. 1984a. Sediment composition and hydrography in six high-gradient estuaries of the northwestern United States. *Journal of Sedimentary Petrology* 54: 86–97.
Peterson, C. D., K. F. Scheidegger, and H. J. Schrader. 1984b. Holocene depositional evolution of a small active-margin estuary of the northwestern United States. *Marine Geology* 59: 51–83.
Phipps, J. B., and J. M. Smith. 1978. Coastal accretion and erosion in southwest Washington. Washington Department of Ecology, Olympia, Report No. WA/DOE/CZ/78-12.
Plummer, K. 1991. *The Shogun's Reluctant Ambassadors.* Portland: Oregon Historical Press.
Price, R. L. 1975. *Newport, Oregon, 1866–1936: A Portrait of a Coast Resort.* Dallas, Ore.: Itemizer-Observer.
Quinn, W. H., D. O. Zopf, K. S. Short, and R. T. W. Kuo Yang. 1978. Historical trends and statistics of the southern oscillations, El Niño, and Indonesian droughts. *Fishery Bulletin* 76: 663–78.
Rankin, D. K. 1983. Holocene geologic history of the Clatsop Plains foredune ridge complex. Master's thesis, Portland State University.
Rea, C. C. 1975. The erosion of Siletz Spit, Oregon. Master's thesis, Oregon State University.
Reilinger, R., and J. Adams. 1982. Geodetic evidence for active landward tilting of the Oregon and Washington Coastal Ranges. *Geophysical Research Letters* 9(4): 401–3.
Rogers, L. C. 1966. *Blue Water 2* lives up to promise. *Oil and Gas Journal,* August 15, 73–75.
Rohleder, J. P. 1981. Environmental hazards report. Phase I, Beachland Estates, Oceanview addition, block 22, Newport, Oregon. Unpublished technical report.
Rohleder, J. P., R. M. Northam, A. J. Kimerling, and C. L. Rosenfeld. 1975. Environmental hazard inventory, coastal Lincoln County, Oregon. Unpublished report.
Ruggiero, P., P. D. Komar, W. G. McDougal, and R. A. Beach. 1996. Extreme water levels, wave runup, and coastal erosion. In *Proceedings of the 25th Coastal Engineering Conference,* American Society of Engineers, forthcoming.
Satake, K., K. Shimazaki, Y. Tsuji, and K. Ueda. 1996. Time and size of a giant earthquake in Cascadia inferred from Japanese tsunami records of January 1700. *Nature* 379:246–49.

Sayre, W. O., and P. D. Komar. 1988. The Jump-Off Joe landslide at Newport, Oregon: History of erosion, development and destruction. *Shore and Beach* 53(3): 15–22.

———. 1989. The construction of homes on four active coastal landslides in Newport, Oregon: Unbelievable but true! In *Coastal Zone '89*, 3286–96. American Society of Civil Engineers.

Schatz, C. 1965. Source and characteristics of the tsunami observed along the coast of the Pacific Northwest on 28 March 1968. Master's thesis, Oregon State University.

Schatz, C. E., H. Curl, and W. V. Burt. 1964. Tsunamis on the Oregon coast. *Ore Bin* 26: 231–32.

Scheidegger, K. F., L. D. Kulm, and E. J. Runge. 1971. Sediment sources and dispersal patterns of Oregon continental shelf sands. *Journal of Sedimentary Petrology* 41: 1112–20.

Scheidegger, K. F., and J. P. Phipps. 1976. Dispersal patterns of sand in Grays Harbor estuary, Washington. *Journal of Sedimentary Petrology* 46: 163–66.

Schlicker, H. G., R. E. Corcoran, and R. G. Bowen. 1961. Geology of the Ecola State Park landslide area, Oregon. *Ore Bin* 23(9): 85–90.

Schwartz, M. L., J. Mahala, and H. S. Bronson. 1985. Net shore-drift along the Pacific coast of Washington State. *Shore and Beach* 53: 21–25.

Seymour, R. J., R. R. Strange, D. R. Cayan, and R. A. Nathan. 1985. Influence of El Niño on California's wave climate. In *Proceedings of the 19th Coastal Engineering Conference,* 577–92. American Society of Civil Engineers.

Shepard, F. P., and J. R. Curray. 1967. Carbon-14 determinations of sea level changes in stable areas. In *Progress in Oceanography.* Vol. 4: *The Quaternary History of the Ocean Basins,* 283–91. Oxford: Pergamon Press.

Shih, S.-M. 1992. Sea-cliff erosion on the Oregon coast: From neotectonics to wave runup. Ph.D. diss., Oregon State University.

Shih, S.-M., and P. D. Komar. 1994. Sediments, beach morphology and sea cliff erosion within an Oregon coast littoral cell. *Journal of Coastal Research* 10: 144–57.

Shih, S.-M., P. D. Komar, K. J. Tillotson, W. G. McDougal, and P. Ruggiero. 1994. Wave run-up and sea-cliff erosion. In *Proceedings of the 23rd Conference on Coastal Engineering,* 2170–84. American Society of Civil Engineers.

Stembridge, J. E. 1975a. *Pathfinder, the First Automobile Trip from Newport to Siletz Bay, Oregon, July 1912.* Newport, Ore.: Lincoln County Historical Society.

———. 1975b. Recent shoreline changes of the Alsea sandspit, Lincoln County, Oregon. *Ore Bin* 37: 77–82.

———. 1975c. Shoreline changes and physiographic hazards on the Oregon coast. Ph.D. diss., University of Oregon.

Straton, K. A. 1977. *Oregon's Beaches—A Birthright Preserved.* Salem: Oregon State Parks and Recreation Branch.

Sunamura, T. 1983. Processes of sea cliff and platform erosion. In *Handbook of Coastal Processes and Erosion,* ed. P. D. Komar, 233–65. Boca Raton, Fla.: CRC Press.

Terich, T. A. 1973. Development and erosion history of Bayocean Spit, Tillamook, Oregon. Ph.D. diss., Oregon State University.

Terich, T. A., and P. D. Komar. 1974. Bayocean Spit, Oregon: History of development and erosional destruction. *Shore and Beach* 42: 3–10.

Terich, T., and T. Levenseller. 1986. The severe erosion of Cape Shoalwater, Washington. *Journal of Coastal Research* 2: 465–77.

Tillotson, K. 1994. Wave climate and storm systems on the Oregon coast. Master's thesis, Oregon State University.

Tillotson, K., and P. D. Komar. 1997. The wave climate of the Pacific Northwest (Oregon & Washington): A comparison of data sources. *Journal of Coastal Research* 13: 440–52.

U.S. Army Corps of Engineers. 1970. Tillamook Bay, Oregon. Committee on Tidal Hydraulics.

Vincent, P. 1989. Geodetic deformation of the Oregon Cascadia margin. Master's thesis, University of Oregon.

Watts, J. S., and R. E. Faulkner. 1968. Designing a drilling rig for severe seas. *Ocean Industry* 3: 28–37.

Weber, B. 1984. *Wrecked Japanese Junks Adrift in the North Pacific Ocean.* Fairfield, Wash.: Ye Galleon Press.

Weber, B., and M. Weber. 1989. *Bayocean, the Oregon Town That Fell into the Sea.* Medford, Ore.: Pacific Northwest Books.

Wilson, B. W., and A. Torum. 1968. The tsunami of the Alaskan earthquake, 1964. Engineering evaluation, U.S. Army Corps of Engineers, Coastal Engineering Research Center, Tech. Memo 25.

Wyrtki, K. 1975. Fluctuations of the dynamic topography in the Pacific Ocean. *Journal of Physical Oceanography* 5: 450–59.

———. 1977. Sea level during the 1972 El Niño. *Journal of Physical Oceanography* 7: 779–87.

———. 1984. The slope of sea level along the equator during the 1982/1983 El Niño. *Journal of Geophysical Research* 89: 10419–24.

Zopf, D. O., H. C. Creech, and W. H. Quinn. 1976. The wavemeter: A land-based system for measuring nearshore ocean waves. *MTS Journal* 10: 19–25.

Index

Paul D. Komar is Professor of Oceanography at Oregon State University. He is author of *Beach Processes and Sedimentation* and editor of *Handbook of Coastal Processes and Erosion.*

Library of Congress Cataloging-in-Publication Data

Komar, Paul D.
The Pacific Northwest Coast : living with the shores of Oregon and Washington / Paul D. Komar.
p. cm.—(Living with the shore)
Includes index.
ISBN 0-8223-2010-X (alk. paper).—ISBN 0-8223-2020-7 (pbk.: alk. paper) 1. Shore protection—Washington (State) 2. Shore protection—Oregon. 3. Coast changes—Washington (State). 4. Coast changes—Oregon. 5. Northwest coast of North America.
I. Title. II. Series TC223.8.K66 1998
627'.58'09797—dc21 97–23870 CIP